Manual of Avionics

An Introduction to the Electronics of Civil Aviation

Third Edition

Brian Kendal

OXFORD
BLACKWELL SCIENTIFIC PUBLICATIONS
LONDON EDINBURGH BOSTON
MELBOURNE PARIS BERLIN VIENNA

Blackwell Scientific Publications
Editorial Offices:
Osney Mead, Oxford OX2 0EL
25 John Street, London WC1N 2BL
23 Ainslie Place, Edinburgh EH3 6AJ
238 Main Street, Cambridge,
Massachusetts 02142, USA
54 University Street, Carlton
Victoria 3053, Australia

Other Editorial Offices:
Librairie Arnette SA
2, rue Casimir-Delavigne
75006 Paris
France

Blackwell Wissenschafts-Verlag
Meinekestrasse 4
D-1000 Berlin 15
Germany

Blackwell MZV
Feldgasse 13
A-1238 Wien
Austria

First published by
Granada Publishing Ltd 1979
Second Edition published by
BSP Professional Books 1987
Reprinted 1989
Third edition published by
Blackwell Scientific Publications 1993

Set by DP Photosetting, Aylesbury, Bucks
Printed and bound in Great Britain by
Hartnolls Ltd, Bodmin, Cornwall

DISTRIBUTORS

Marston Book Services Ltd
PO Box 87
Oxford OX2 0DT
(*Orders:* Tel: 0865 791155
Fax: 0865 791927
Telex: 837515)

USA
Blackwell Scientific Publications, Inc.
238 Main Street
Cambridge, MA 02142
(*Orders:* Tel: 800 759-6102
617 225-0401)

Canada
Oxford University Press
70 Wynford Drive
Don Mills
Ontario M3C 1J9
(*Orders:* Tel: 416 441-2941)

Australia
Blackwell Scientific Publications Pty Ltd
54 University Street
Carlton, Victoria 3053
(*Orders:* Tel: 03 347-5552)

British Library
Cataloguing in Publication Data

A catalogue record for this book is available from the British Library

ISBN 0-632-03472-6

Library of Congress
Cataloging in Publication Data

Kendal, Brian.
Manual of avionics: an introduction to the electronics of civil aviation / Brian Kendal.
p. cm.
Includes index.
ISBN 0-632-03472-6
1. Avionics. I. Title.
TL695.K46 1993
629.135—dc20 92-28100
CIP

Contents

Preface

Many people express a wish to write a book, but few are offered the opportunity to proceed further. I therefore owe a great debt to my colleague Ronald Hurst for not only suggesting that I should put pen to paper but also arranging the necessary introductions which made publication possible.

Once afforded such an opportunity, the prospective author must decide for whom the book is intended. Memories of the perusal of technical books over a period of over thirty years give the impression that, in general, such books are written at one of two levels – either in an extremely elementary form or alternatively as a complex mathematical analysis of the subject. I therefore decided to attempt to steer a course somewhere between the two and, in so doing, hopefully produce a volume which could be of both interest and utility to those concerned with the subject of civil aviation telecommunications and radio aids to navigation.

I have included descriptions of many systems which are no longer representative of modern technology. The reason for this is twofold. First, a belief that by knowing the line of development from past to present, it is far easier to comprehend that from present to future. Second, airport systems are built to extremely high standards and are consequently very expensive. It is not surprising, therefore, that in various, less affluent, parts of the world, older equipment is still in operation, meeting the original specifications and still performing an excellent job.

The pace of development in the field of aviation electronics ensures that any volume concerned with the subject must be updated at relatively frequent intervals. For example, practical space systems have been developed since the first edition went to press and digital recording systems for both audio and radar have been introduced since the second.

Political events also have a considerable effect. For example, the political changes in what was formerly the Soviet Union have made it possible to describe, in this third edition, the Glonass satellite navigation system, about which little, other than its existence, was known just a few years ago.

Undoubtedly, the development of satellite navigation systems using either or both Navstar and Glonass will be the most important feature of the next decade, for not only have they global coverage, but their accuracy is such that they may well render unnecessary many of the CAT 1 landing systems currently used in many parts of the world.

Even closer to home, it had been expected that the Decca Navigator system would cease operations in the early 1990s, but political decisions have ensured that it will be still current at the turn of the century and perhaps for much longer. For this, I am able to describe a new generation of Decca Navigator Equipment which is fully compatible with modern navigational computers.

Brian Kendal
May 1992

Acknowledgements

The wide ranging nature of a book such as this inevitably necessitates description of subjects outside the author's personal experience. On these topics one must rely on the generosity and expertise of others. For this reason I would like to express my grateful thanks to the many organisations, colleagues and friends who so willingly gave me the benefit of their expertise and experience. In this context I should particularly like to mention my friends Harry Cole (of Marconi Radar) and Walter Blanchard, without whose assistance the revision of this book would not have been possible.

I should also like to acknowledge the assistance which I have received from the following firms:

Aeronautical and General Instruments Ltd; Civil Aviation Authority; Becker Flugfunk; Collins; Decca Navigation Co.; Fernau Electronics; Litton Systems (Canada) Ltd; Marconi Radar Systems Ltd; Marconi Secure Radio Systems Ltd; Park Air Electronics Ltd; Plessey Radar Ltd; Racal Avionics Ltd; S. E. Labs (EMI) Ltd; Standard Elektrik Lorenz AG and Walton Radar Systems Ltd.

Section 1

A brief history

1.1 The origins of radio

In his book *Treatise on Electricity and Magnetism*, published in 1873, James Clerke Maxwell brought together the known facts concerning light, electricity and magnetism. Developing from his postulation of the electromagnetic theory of light was the prediction that other waves existed which would propagate through space with a velocity equal to that of light. These would be produced whenever oscillatory currents were set up and would obey the classical laws of geometric optics.

In 1879 Professor D. E. Hughes of London demonstrated to a group of distinguished scientists that it was possible to transmit signals over several hundred yards without the use of interconnecting wires. His first experiments were conducted in his own home but on later occasions he walked up and down Great Portland Street with a telephone receiver to his ear, hearing signals up to a distance of almost 500 yards from the transmitter. For these demonstrations he used an induction coil for the transmitter and a microphonic joint with a telephone earpiece as a receiver. It is interesting to note that much of Professor Hughes' early apparatus now has a permanent home in the Science Museum in London.

Some five years later a Professor Onesti demonstrated that if iron filings were placed in a tube of insulating material between copper electrodes, the application of a fairly high voltage could cause them to cohere, or stick together sufficiently to allow a current to pass. Revolving the tube decohered them.

In 1889 Oliver Lodge again demonstrated the same phenomena, this time between two metal spheres, and later manufactured a coherer using a microphonic contact between a watch spring and an aluminium plate. Some two years later Professor Branley verified the previous experiments and also demonstrated that the filings could be made to cohere by an electrical discharge in the vicinity of, but not connected to, the coherer. Oliver Lodge recognised the importance of Branley's coherer and further improved it by adding a mechanical tapper to return the filings to a non-conductive condition. Using this apparatus he demonstrated equipment with a range of 150 yards to the British Association but failed to realise the potential of the device.

Meanwhile the existence of these waves was also experimentally verified in 1887–8 by Heinrich Hertz whose further work was concerned with proving that these waves did have the quasi-optical properties predicted by Maxwell fifteen years previously.

In his experiments Hertz used spark-gap transmitters operating initially on a wavelength of 10 m and later on 66 cm. In each case, the transmitter was placed along the focal line of a cylindrical reflector and the receiver, which consisted of a spark gap at the centre of a dipole, was situated along the centre line of a further similar reflector. With this simple apparatus, Hertz was able to detect radiation at a distance of several metres and by rotating the receiver by 90° showed that the radiations were polarised. This was further confirmed by an experiment in which he interposed a wire grating between transmitter and receiver and demonstrated that if the wires of the grating were parallel to the transmitting and receiving dipoles, the grating was transparent to the radiation but if at 90° it was opaque. He also demonstrated shadowing by opaque objects and by means of a large prism of pitch cast in a wooden box, he showed that refraction as much as 22° was possible. From these experiments he deduced that the Refractive Index of pitch to these electromagnetic waves was 1.69 compared with a value of between 1.5 and 1.6 for similar substances when transmitting light.

In the succeeding decade many workers extended Hertz's work using even shorter wavelengths. With wavelengths in what is now known as the centimetric region, quasi-optical experiments were performed which would not have been possible with Hertz's metric wavelengths. Typical of these was a demonstration of double refraction of 8 mm waves by Peter Lebedew in Moscow.

In 1895, Popov in Russia, applying Hertzian principles to the study of atmospheric electricity, developed a receiver for the reception of Hertzian waves which worked quite well over limited distances. In the same year, however, Marconi produced the first really reliable detector using a coherer of his own invention.

In his first practical demonstrations on Salisbury Plain the following year, Marconi used a wavelength of 1 m, utilising parabolic reflectors behind both transmitter and receiver. Later, however, due to ease at which they could be generated, Marconi's interest turned to longer wavelengths and it was not for nearly two decades that he again returned to the metric wavelengths. Also in 1895, a naval officer, Captain H. B. Jackson succeeded in establishing ship-to-shore wireless communication, but for military reasons the work was carried out in secret and the results were never published.

In 1896 Marconi patented a wireless system based on the work of Hertz some eight years previously, and embarked on a series of tests over gradually increasing distances and in 1897 succeeded in working over a distance of $8\frac{3}{4}$ miles. Over-water experiments were also carried out between Lavernock Point near Barry in Glamorgan and the Island of Steep Holme in the Bristol Channel.

In 1898 contact was established between Dover Town Hall and Wimereux in France and this was followed by direct communication between Wimereux and the Marconi Factory in Chelmsford–a distance of 85 miles.

In December 1901 signals from the Marconi high power wireless station at Poldhu were heard in St Johns, Newfoundland, and from that time onwards wireless became recognised as a long distance communications medium.

1.2 Towards ILS and VOR

The ancestry of the currently used landing and en-route aids (ILS and VOR) can be traced back to the earliest days of radio communication. Their lines of descent cross in many places, but both ultimately find their origins in two patents filed in Germany in 1906.

By the year 1905 Marconi had expended considerable effort in the investigation of the properties of the classic inverted 'L' aerial. He found that if the horizontal limb were considerably longer than the vertical section, the polar diagram exhibited a considerable bulge in the opposite direction to the line of the horizontal section.

In 1905 he patented a system which used these inverted 'L' aerials for both transmitting and receiving, claiming exceptional directional properties for the combination. Expanding on the same principle, Marconi filed a patent the following year for an aerial system of a number of inverted 'L' aerials evenly spaced radially around the receiver. By selecting the aerial receiving the strongest signal, the approximate direction of the transmitting station could be ascertained.

Before twelve months had elapsed, Telefunken of Germany had introduced a very similar idea but this time in the transmitting idiom. This consisted of a transmitter which radiated first a prearranged 'start' signal to a central omni-directional aerial followed by one second transmissions on each of thirty-two aerials spaced radially around the central omni-directional radiator at each of the points of the compass.

A station wishing to use the beacon merely had to start a stopwatch on hearing the 'start' signal and stop it when the signal reached its maximum strength. To assist observers a special watch was issued. This had a hand which made one revolution in thirty-two seconds and the face was calibrated in the points of the compass.

Although this system never came into wide use, it may well be considered as the forerunner of all modern rotating beacons.

In 1907 Bellini and Tosi produced a design comprising two loop receiving aerials crossed at right angles from which the directions of incoming waves could be determined from the relative magnitude of the currents that they induced in the aerials. This was developed further and throughout the 1914–1918 war considerable reliance was placed on the Bellini-Tosi system by both major antagonists. Despite the fact that any one chain of direction-finding stations could handle only one aircraft at a time and also that unsuspected propagation phenomena sometimes resulted in positional errors in excess of fifty miles, von Buttlar-Brandenfels, the only Zeppelin

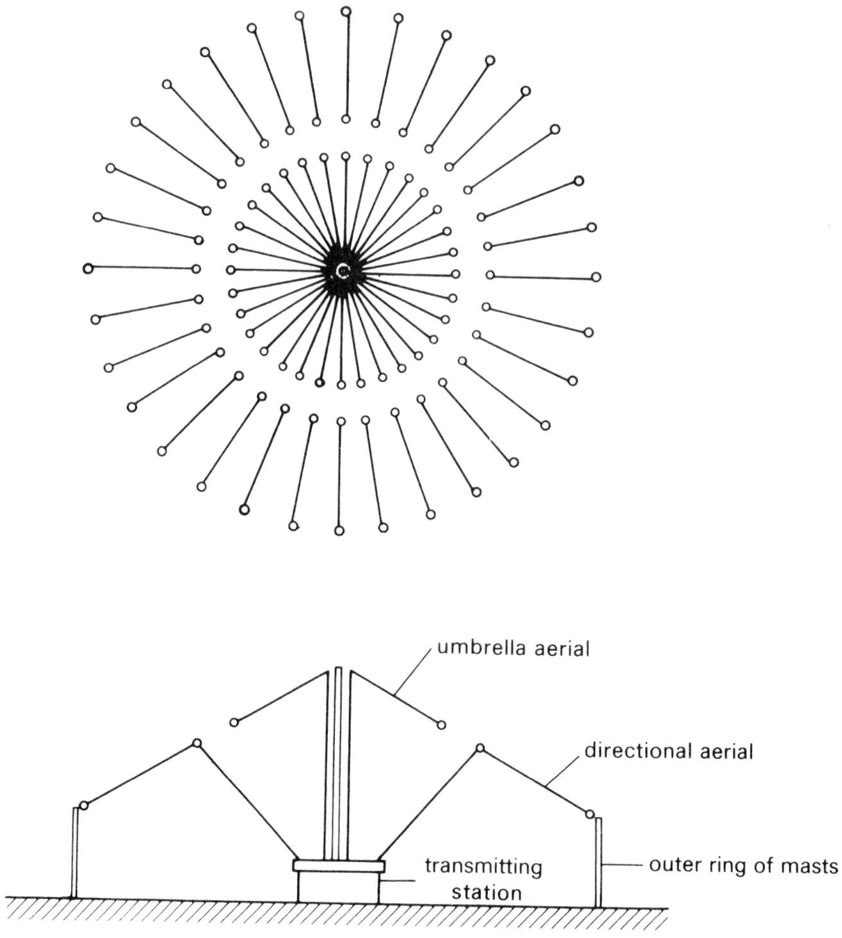

Fig. 1 The Telefunken compass

commander who flew throughout the war, concluded that radio navigation was far superior to celestial.

The disadvantages of reliance on a D/F system for navigational purposes are that the D/F station is rapidly overloaded and that in wartime the transmission of an aircraft reveals his position to friend and foe alike. Nevertheless D/F systems using the Bellini-Tosi or a later system developed by Adcock remained a vital part of air navigation for the next thirty years.

From 1916 Marconi became involved with 'short wave directional wireless' in a series of trials which were held variously at Hendon and Caernarvon and from these he developed the 'wireless lighthouse' which was installed on Inchkeith Island in 1921.

Probably the most important fact realised by Marconi in the design of this equipment was that only by raising the frequency up to the VHF spectrum could a sufficiently sharp beam be generated to give worthwhile accuracy. Furthermore the design of the keying arrangements were such that it was impossible for the beacon to radiate inaccurate information and the use of a stopwatch was also made unnecessary.

Operating on a wavelength of 6.2 m the Inchkeith beacon aerial consisted of two back-to-back paraboloids of 13 m aperture, each fed from a separate transmitter. Around the base of the aerial, on a large ring, contact segments were arranged in morse characters, which operated a contact breaker to key the transmitters as the aerial assembly revolved, transmitting a morse letter at each bearing. Between these letters corresponding to bearing, the station identification letters were transmitted. The aerial array rotated once every two minutes, the back-to-back aerials permitting the receiving operator to take a bearing once every minute.

On board the receiving ship, once the signal had been tuned, the wireless operator had only to reduce the gain of the receiver until only just one or two letters could be heard, these corresponding to the relative bearing between Inchkeith Island and the ship.

The principle must have been successful for a further beacon was installed at South Foreland in 1926. This had a similar radiation pattern but used a broadside aerial array, 76 ft long by 30 ft high designed by Franklin.

Concurrently with Marconi's work, the Wireless Dept of the Royal Aircraft Establishment developed and tested an MF rotating beacon running 500 watts to a six-turn loop, 5 ft square operating on a frequency of 500 kHz. A rotating beacon was also installed at Orfordness. This used a complex keying sequence which gave both north and west starting signals.

Returning to 1907, a patent by O. Scheller of the Lorenz Company led to a line of development which, when eventually combined with the rotating beacon, culminated in the aids that we know today.

This was the 'Course Setter', the principle of which was that two aerials with intersecting radiation patterns were energised alternately from a single transmitter. The power fed to one aerial was keyed to form the letter 'A' so in complement the power to the other aerial keyed 'N'. The position of the aerials was arranged such that the intersection of the radiation patterns corresponded with the desired course. Therefore, if the ship or aircraft were off course either an A or N would be heard but when on course both radiations would be heard simultaneously combining into a steady tone.

In 1917 tests were made in Germany using this system with ships and five years later Keibitz arranged a further series of trials using aircraft. The aerials then used were 140 m long, intersecting at 20° and operating at wavelengths of 350 m and 550 m. Although on the ground accuracies of 30 m were claimed at a range of 3.5 km, the results obtained in the air were confusing and development ceased until the early 1930s when the company again used the equi-signal principle in their VHF beam approach equipment.

The Lorenz beam approach equipment consisted of a VHF transmitter situated at the upwind end of the runway operating at a frequency of approximately 33 MHz. This fed to a vertical dipole on either side of which, at right angles to the runway, was positioned a reflector element. The centre point of each of these reflector elements was broken and bridged by a set of relay contacts which operated in opposition, thus if one set were 'made' the other would be 'broken', rendering that reflector inoperative. By keying the relays the radiation pattern could be moved from side to side. The two patterns intersected on the line of the runway. The keying of the reflectors was arranged to be such that one reflector was 'made' for three times the period of the other, thus when approaching the runway slightly off course, the pilot heard a series of either Es or Ts which merged into a steady tone as the correct alignment was achieved.

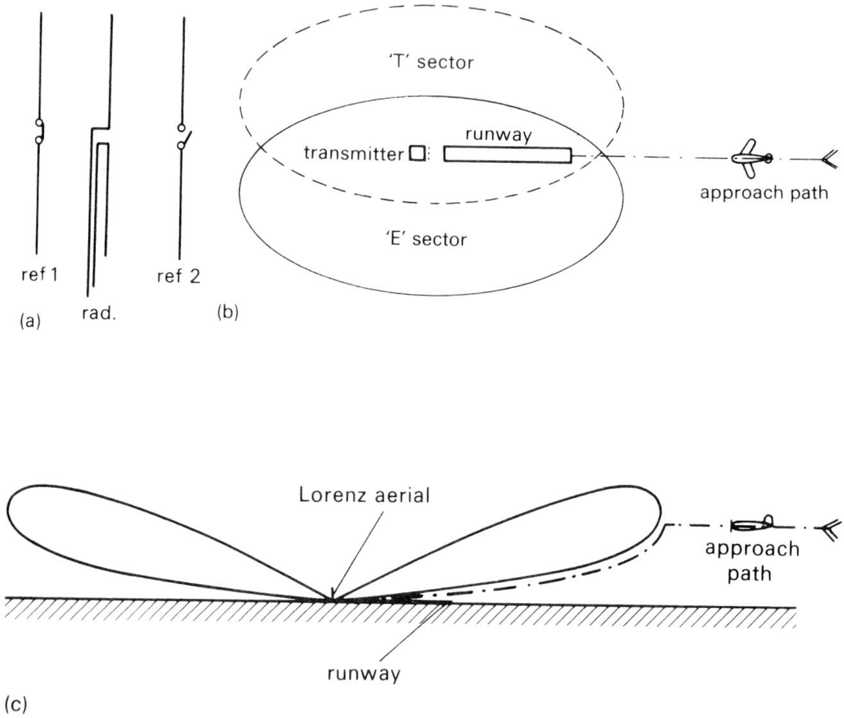

Fig. 2 Lorenz beam approach. (a) Lorenz aerial arrangement. (b) Radiation pattern. (c) Lorenz vertical polar diagram and approach path.

The aircraft receiver was fitted with a signal strength meter and glide slope guidance could be obtained by following a contour of equal field strength.

This system was also adopted in the UK where it was known as standard beam approach; it was extremely successful and remained in service until the early 1960s.

Meanwhile, in the USA, development was proceeding on a system of feeding interlocking signals to crossed loop aerials. This method generated four separate courses from each station and a course width of 500 ft at 50 miles was claimed but course distortions were observed due to reflections from railway lines etc.

From 1923 to 1926 the USA Army continued with this line of development and found that by using a modified transmitting goniometer in conjunction with Bellini-Tosi loops, the courses could be moved to almost any desired direction. The success of this system was such that in 1926 the newly formed Aeronautics Branch of the Department of Commerce undertook to install a system of these beacons to delineate the increasing number of air routes within the USA. The Bellini-Tosi aerial systems were found to be susceptible at night to errors of up to 40° under adverse conditions but a change to Adcock aerials alleviated this problem. By 1944 over 300 of these 'Radio Ranges' had been installed and remained in service until superseded by VOR.

A further development with the Radio Range transmitter was the abandonment of the interlocking signal and its substitution by tones of 65 Hz and 87.5 Hz. After detection in the aircraft receiver the tones were separated by filters and the two outputs fed differentially to a centre-zero meter such that if the aircraft was on course the meter indicated zero with deflections to left or right when off course.

In one experiment such a transmitter, operating on 330 kHz, was positioned at the upwind end of a runway with one course aligned directly along the runway. Additionally a 93.7 MHz signal provided vertical guidance in the same way as the Lorenz system. On 15 September 1931, pilot M. S. Boggs made the first of over 100 blind landings using this system on a runway measuring only 2000 ft by 100 ft.

In 1938 several interested parties co-operated in the preparation of a report which outlined the basis of Instrument Landing System (ILS) as we know it today.

The main recommendations were that:

(a) the system should operated in the 108 MHz to 112 MHz band
(b) the vertical guidance path should be a straight line
(c) two distance markers should be installed operating on 75 MHz.

This specification led to the development of the SCS 51 equipment, the forerunner of today's ILS. In this, only two variations from the previous experiments were made: the modulation frequencies were changed to 90 Hz and 150 Hz and the vertical guidance frequency to 330 MHz. Although the techniques used in ILS systems have been continually refined since that time, the basic system parameters set by the SCS 51 equipment have never changed.

Consequent to the problems of operating medium frequency radio ranges (night effect etc.), in 1937 the U.S. Civil Aeronautics Administration conducted a series of tests involving the use of VHF for radio range beacons. Early tests used a frequency of 63 MHz and results were promising, but problems arose due to reflection effects under abnormal propagation conditions and in consequence the operational frequency was increased to 125 MHz. Although some problems still remained with the four-course systems, the work on VHF had shown that equipment operating on these frequencies was capable of better performance than on MF. A major problem remaining however was the disorientation experienced by a pilot losing his bearings within a four-course range; the viability of a two-course range system was therefore investigated. The result of this was the development of the Visual-Aural Two Course range in which two patterns were radiated simultaneously, delineating four separate courses. This could comprise typically of an east-west course delineated by tones of 150 Hz and 90 Hz to the north and south respectively, these tones being separated by filters in the aircraft receiver and fed differentially to a centre-zero meter. This was known as the visual course. Additionally a north-south pair of courses was keyed at right angles using interlocking (D and U) Lorenz-type signals.

By 1936 consideration was also being given to a beacon radiating an infinite number of courses, this being in essence a revival of the rotating beacon.

This beacon employed a system in which the horizontal polar diagram was a limaçon, this pattern having the property that if rotated, the signal strength at any receiving station varies sinusoidally. This pattern was rotated at a constant 60 Hz. The received sine wave signal was split into two parts in phase quadrature, which were applied to the deflection plates of a cathode ray tube. This produced a circular trace, each point of which was associated with the time corresponding to some particular orientation of the space pattern but without a reference point. This was originally provided by introducing a break in the signal as the maximum of the rotating pattern passed through true north, the effect being to make a radial deflection on the otherwise circular trace and thus giving an indication of the bearing.

In the early work on this beacon, a frequency of 6.5 MHz was used, but in view of the tendency in favour of VHF, tests on this frequency were discontinued and further development used frequencies in the 125 MHz band.

A later version of the equipment substituted a further modulation for the momentary reference break, this being a 60 Hz modulation applied to a 10 kHz subcarrier whose phase was arranged to coincide with the rotating pattern at true north.

In the receiver, the phases of the reference and rotating signals were compared, the phase difference corresponding to the bearing of the receiver. Consequent to this change in reference signal, the use of the cathode ray tube indicator was discontinued and was replaced by a phase meter type of azimuth indication.

This was, in essence, the VOR in use today, the main subsequent variations being a change in frequency to the 112.0 MHz to 117.9 MHz band, reduction of pattern rotation speed and reference modulation to 30 Hz and the selection of 9960 Hz for the subcarrier frequency.

1.3 Towards radar

In 1922 Marconi, speaking as a guest of honour at a joint meeting of the Institute of Electrical Engineers and the Institute of Radio Engineers in New York, said that it seemed to him that it should be possible to design an apparatus by means of which a ship could radiate a divergent beam of rays in any desired direction. If these rays impinged on a metallic object such as another ship, they would be reflected back to a receiver on the sending ship, thereby immediately revealing the presence and bearing of the other ship regardless of the visual conditions at the time.

Marconi was not the first to consider this possibility, for in 1904 an engineer from Düsseldorf, Christian Hulsmeyer, patented the idea in several countries. A prototype equipment, which was called the telemobiloscope, was constructed and demonstrated from a Rhine bridge at Cologne, detecting an approaching barge.

Later that year, Hulsmeyer's apparatus was installed in the tender *Columbus* which cruised up and down Rotterdam harbour detecting vessels up to 5 km in range. Unfortunately, even these demonstrations were insufficient to encourage any orders.

The Hulsmeyer apparatus embodied many very advanced ideas. The operating wavelength was 50 cm, with the receiving and transmitting antennae effectively screened from each other. The latter was a parabolic reflector with additional stacked director elements, 25 years before Professor Yagi published his work on the antenna which bears his name.

Unfortunately, as demonstrated, the telemobiloscope was only capable of determining the presence and bearing of the target, although Hulsmeyer later worked on a ranging system using triangulation techniques.

Despite all efforts, no sales were ever made and Hulsmeyer eventually turned to other work.

An experimental confirmation of the practicality of these proposals was achieved in the autumn of 1922 when A. Hoyt Taylor and Leo Clifford were performing an informal investigation of 5 m waves at the United States Naval Research Laboratory at Anacostia, D.C.

A receiver was installed in a motor car and a transmitter set up adjacent to the door of their laboratory. As the car was driven away from the laboratory, interference effects were noticed almost at once, for as the car passed certain steel buildings, signal strengths fluctuated wildly and passing cars and networks of wires such as those surrounding tennis courts, caused pronounced shielding effects. Whilst investigating the propagation of these wavelengths across water, the car containing the receiving equipment was driven to a site across the Potomac River where interference effects were immediately noticed. These were found to emanate from clumps of willow trees near to the receiver. When a small steamer, the *Dorchester*, passed on its journey down the channel, an interesting effect was recorded. Fifty feet before the bow of the steamer intersected the line between transmitter and receiver, the signal strength jumped to nearly twice the normal value. As the steamer crossed the line, signals dropped to half strength but when the steamer has passed and the stern was 50 ft past the line, signals again doubled in strength before returning to the normal level.

This phenomenon led to a suggestion by Taylor that the effect might be utilised to detect the passage of an enemy vessel through a line of warships, sailing in line abreast formation, several miles apart, regardless of weather conditions. This is probably the earliest suggestion for the use of radio detection for military purposes.

It was not until 1930 that the subject of the radio detection of objects was again broached, this time by an Englishman, W. J. Brown, who wrote in the 'Proceedings of the Institute of Radio Engineers' that although it had already been suggested that icebergs could be detected by short wave radio, he considered that a similar method might be used for estimating the height of an aircraft above ground.

Ranging techniques

Long before the advent of radio, efforts were being made to measure distance by timing of reflected pulses of energy. It appears that the technique was first suggested by the French physicist Arago about the year 1807. Early experiments concerned with the measurement of the depth of oceans used a method which involved the detonation of a charge of gunpowder on the bottom of the sea and determining the interval before the sound was heard. Even as late as 1912 this technique was successfully employed by Behn in the determination of the depth of Lake Ploen.

The British Admiralty developed a somewhat more sophisticated application of this technique using a steel hammer striking a steel plate in the bottom of a ship. This generated a highly damped compression wave which after reflection from the sea bed was received by hydrophones. A similar principle was used by Fessenden, who used a transmitter which emitted a short pulse of energy on a frequency of a few thousand cycles per second. Lewis Richardson proposed that high frequency supersonic beams could be used for both depth measurement and obstacle detection at sea and this became practical after Langevin and Chilowsky produced beams at 30000 Hz to 40000 Hz. Before the end of the First World War the USA had developed a system which could detect submarines at a range of half a mile. Succeeding developments of this system became the sonar equipment of the Second World War.

After Marconi's success in receiving signals across the Atlantic, in

Plate 1 The actor Robert Loraine using experimental Marconi aircraft communication equipment on Salisbury Plain, 1910. *(Photo Marconi)*

Plate 2 The Marconi type AD2 aircraft wireless telephony equipment produced in the 1920s – the world's first commercial aircraft equipment *(Photo Marconi)*

Plate 3 Marconi Belli-Tosi direction finding receiver and aerodrome transmitter remote control equipment at a temporary site before the Croydon airport buildings were constructed. *(Photo Marconi)*

Plate 4 Croydon Airport control tower surmounted by the Bellini-Tosi direction finding loops. The aircraft is the Handley Page HP42 Horatius. *(Photo Marconi)*

Plate 5 The transmitter room of a Chain Home (CH) radar station. *(Photo Marconi)*

Plate 6 The aerial system for CH radar. The transmitting array was suspended between the 360 foot masts on the left and the receiver aerials were attached to the 150 foot masts on the right. *(Photo: Marconi)*

December 1901, several workers in the field put forward theories as to the mechanism which made such communication possible. Working independently on opposite sides of the Atlantic, both Arthur Edwin Kennelly in America and Oliver Heaviside in England postulated that the signals were propagated around the Earth with the assistance of a conducting surface in the upper atmosphere. Neither worker attempted to offer any serious explanation as to its nature and it was left to Eccles to suggest, in 1912, that the waves were refracted by ions in the layer. In the succeeding years evidence of various kinds accumulated in support of the theory of an ionised layer but it was not until 1924 that its existence was systematically verified.

Early in that year, John Reinartz, a well known radio amateur, began a series of tests on 20 m and 60 m and noticed that the signals, in contrast to those of longer wavelengths, become weaker after sunset but nevertheless were audible at great distances. The ground wave petered out after a short distance but after a 'dead belt' the signals were again audible. This phenomenon was confirmed by A. Hoyt Taylor who called it the 'skip' or 'miss' region.

The scene now moved to England, where, in 1924 E. V. Appleton and M. A. F. Barnett believed that for wavelengths of 300 m to 500 m there would be a point about 100 miles from the transmitter where the ground and sky waves would be of comparable strength and strong interference effects would be produced. This was confirmed experimentally on the nights of 11 December

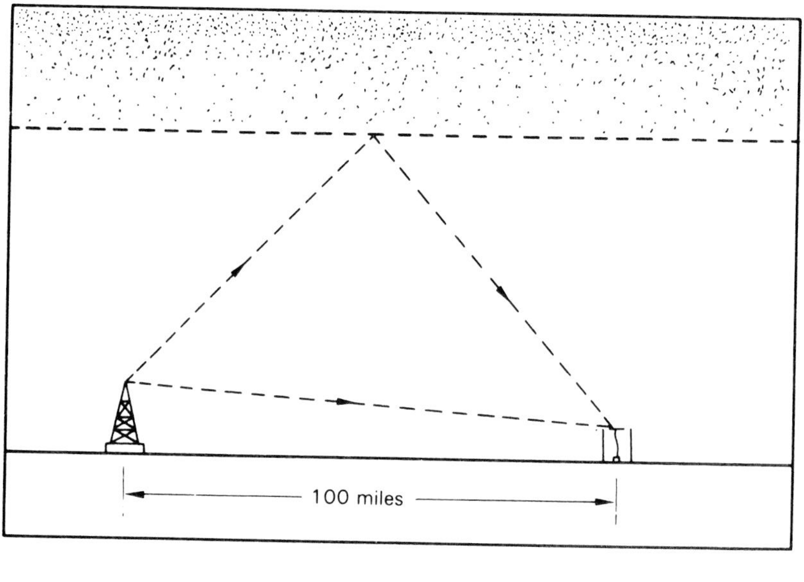

Fig. 3 The Appleton-Barnett experiment

1924 and 17 February 1925 when, using the BBC transmitter at Bournemouth, they obtained interference effects which were the first direct evidence of the reflecting layer which they estimated to be at a height of about 80 km to 90 km.

Unaware of this work, and at about the same time, two American investigators, Tuve and Breit, developed the more simple technique of transmitting pulses and measuring the time for the returning echo to reach the receiver.

Using the Naval Research Laboratory transmitter, operating on a wavelength of 71.3 m, transmitting pulses of 1/1000th second duration, Tuve and Breit estimated the height of the ionosphere to be between 50 and 130 miles. In this work the received pulses were recorded by an oil-immersed mechanical oscillograph and observed by means of a rotating mirror.

The methods of these two workers were widely adopted in the succeeding years by investigators all over the world, resulting in an increasingly detailed knowledge of the upper atmosphere. This was found to comprise, not a single layer as at first thought, but of two or three layers, depending on whether it was day or night time.

Technique was markedly improved when the mechanical oscilloscope was replaced by the cathode ray tube. This method was used in 1930 by Georg Goubau who used a circular trace, and a year later Appleton adopted a system whereby a linear trace was used for his ionospheric experiments.

From the knowledge acquired in these ionospheric investigations, techniques became available that led to the simultaneous development of radar in several countries.

1.4 The development of radar in the UK and Germany

In the early 1930s wild stories were circulating within the British popular press regarding the existence of a 'death ray'. Although this concept seemed far-fetched, the then Director of Scientific Research at the Air Ministry, Dr H. E. Wimperis felt it necessary to assess the feasibility of such a proposal. He therefore approached Robert Watson-Watt who at that time was leading a group of scientists at the Radio Research Station at Slough.

A few calculations by a member of Watson-Watt's staff, Arnold Wilkins, sufficed to prove that energy considerations made such a ray impractical. However, in response to a request from Watson-Watt for other suggestions which might assist the Air Ministry, Wilkins remembered hearing that an aircraft flying in the path of an early experimental VHF radio link had caused severe signal strength fluctuations and suggested that this effect might form the basis of a means of aircraft detection.

Further calculations suggested that at a wavelength of about 50 m, the 75 ft wingspan of a typical RAF heavy bomber of the time should resonate as a half wave dipole aerial thus forming an excellent reflector of radio energy.

Watson-Watt pursued this theme of radio detection of aircraft, proposing a pulse technique to measure range, bearing and height and in addition a system of IFF (Identification Friend or Foe).

In consequence of these suggestions, the Air Ministry requested a trial which was held on the 26 February 1935.

A demonstration was devised whereby one of the BBC's Daventry transmitters radiated a signal on 49 m. A mile away from this transmitter a van was equipped with a receiver incorporating a cathode ray oscillograph. An RAF Handley Page Heyford bomber from Farnborough was detailed to fly a course which crossed the path between the Daventry transmitter and the van.

As the aircraft approached, severe fluctuations of signal strength were seen, which only subsided when the aircraft was eight miles away. This effect was due to the reflected signal from the aircraft combining with the direct signal in various phase relationships, adding to and subtracting from the signal received from the BBC transmitter on the direct path.

Encouraged by these results, the Air Ministry financed a small team of scientists, led by Watson-Watt himself, to investigate radio detection of aircraft. In May 1935 this team moved to a site near Orfordness in Suffolk. A 70 ft wooden mast was erected and, drawing on past experience in ionospheric research, a pulse transmitter and a suitable receiver installed. The aerial system was arranged to radiate a 'floodlight' of energy across the North Sea.

Three days after arrival, returns were being received from the ionosphere and by early June it was possible to demonstrate the detection of aircraft up to a range of 17 miles. Within two months, heights were also being determined by comparing the signals received by two aerials at different heights on the mast. The final problem, that of determination of bearing, was solved in January 1936 by using a direction finding technique based on a pair of dipole aerials mounted at right angles.

The success of the work at Orfordness led to the establishment of a chain of such stations around the British Isles. Operating on frequencies between 22 MHz and 30 MHz with a power of 200 kW and a maximum range of 120 miles, these were known as Chain Home radars (CH).

The Chain Home System exhibited one grave deficiency–it could not detect low-flying aircraft. This problem had been anticipated as early as 1936 and led to the production of a further equipment, Chain Home Low (CHL), which operated on a frequency of 200 MHz. These were mounted on 185 ft towers and had an effective range of fifty miles.

The aircraft plots obtained from the CH and CHL chains could be passed by telephone to centralised operations rooms to be replotted on a large map on which a Fighter Controller could watch the development of a battle and deploy his forces accordingly. Such was the state of development when war was declared on 3 September 1939.

German pre-war development

In the early 1930s, the German Navy's Signals Research Department was developing devices for underwater detection using pulse techniques. In early 1933 it occurred to the head of the establishment, Dr Rudolf Kuhnold, that a similar principle could possibly be applied to radio waves for surface detection. Working towards this goal, early experiments used continuous waves, in effect rediscovering Hulsmeyer's 'Telemobiloscope' of 1904. Within a year Kuhnold progressed to using pulsed transmissions and measuring the time lapse between transmission and reception. From this the distance of the reflecting object could be calculated.

By 1935 a naval set was working on 600 MHz capable of picking up a coast-line at 12 miles and other ships at 5 miles. In early 1936 the frequency was reduced to 150 MHz and after further development this equipment evolved into the 'Freya' radar with a range of 75 miles. This remained in extensive use until the middle of the Second World War.

A higher frequency set operating on 375 MHz known as 'Seetakt', was also manufactured for naval use.

The firm of Telefunken entered the field in 1936, producing by 1938 a set, designated 'Wurtzburg', which worked on 560 MHz. This gave great accuracy but was limited to a range of twenty-five miles.

The magnetron

The development of high accuracy radar had been seriously retarded by the inability of scientists to develop a method of producing high RF power levels at frequencies above 500 MHz. This problem had been concerning a group of scientists at Birmingham University who were mainly concentrating on developing the Klystron which had originally been designed in the United States. This, however, only generated low power oscillations.

Two members of the team, John Randall (now Sir John Randall) and Harry Boot came to the conclusion that the Klystron would never generate the power required and started exploring other avenues. Among these was the split-anode magnetron, a device which, although capable of producing output on centimetric wavelengths had proved to be too frequency unstable for practical use.

The Klystron uses two resonant cavities, known as Rhumbatrons, which at that time were thought to be necessary in the design of any centimetric valve. Randall and Boot therefore decided to combine such cavities with the magnetron principle, i.e. the cavity magnetron.

The basic principles of the cavity magnetron is that an emitter of electrons, i.e. a cathode, is located in the centre of a circular solid metal anode in which a number of cavities, resonating on the required frequency, are machined. On application of a d.c. voltage between cathode and anode, electrons will flow on a straight path across the intervening space. If however the valve is placed in the field of a powerful magnet, the electrons take a circular path. Under suitable conditions the passage of the electrons past the machined slots would cause an oscillation to be set up.

On 21 February 1940 the prototype was ready for testing and when power was applied, Randall and Boot were amazed to find that the output was no less than 400 watts.

This was, of course, a laboratory experiment and was in no way suitable for operational use. However, the data obtained was passed to the Hirst Research Laboratory at Wembley who, within a period of only three months, designed and manufactured a small number of production prototypes.

In August one of these was supplied to the United States by the Tizard Mission and within a few months, the combined production capacity of the United States and the UK ensured that sufficient magnetrons were available for all operational purposes.

1.5 The war years

The threat of war inevitably leads to a considerable increase in the rate of technical development. During hostilities, scientists of the opposing factions are continually engaged in developing new technologies, either as countermeasures to those developed by the enemy or to replace those which the enemy has succeeded in countering. The full story of the developments in radio techniques in the Second World War includes elements of espionage, photo-interpretation, commando raids and intelligence work in addition to scientific development and investigation of the highest standard. So successful were these techniques that in some cases the countermeasures were operational before a particular development could achieve full operational status. Such a story is outside the scope of this book, but a brief description of some of the principal developments in the fields of navigational aids and radar may prove of interest.

The German beams

In the early 1930s, the Lorenz company developed a bad weather approach system based on interlocking morse signals radiated such that when to one side of the approach path the pilot heard a series of dots, a series of dashes on the other and a continuous tone when on the correct course. This aid was used extensively by civil and military organisations in several countries including both the RAF and the German Air Force (the Luftwaffe). Developed further, it became the basis of the Luftwaffe's standard bombing aids.

Knickebein

This comprised two high power Lorenz-type transmitters, located a considerable distance apart, aligned such that the beams intersected over the intended target. The frequencies used were the same as those used for blind approach thus giving the advantage that no special airborne equipment was necessary. The equipment was capable of providing a beam only one third of a degree wide, resulting in an accuracy of 1 mile at a distance of 180 miles.

X-Gerät (Wotan I)

Devised by a specialist in radio propagation, Dr Hans Plendl, this system again made use of a Lorenz-type transmission. The system comprised a main navigational beam intersected by further beams at distances of 30, 12 and 3 miles from the target. The operational frequencies were between 66 MHz and 75 MHz.

The bomber approached the target along the main navigational beam and on hearing the first of the cross beams the pilot knew that the aircraft was accurately aligned on course. When the second beam was heard, the navigator pressed a button to start a special clock. This was not unlike a stop watch except that it had three hands, red, green and black. On the depression of the activating button, the green and black hands began moving together, the black following the green by a time corresponding to the length of the horizontal plot of the bomb's trajectory. On intersecting the third cross beam the button was again depressed which stopped the moving hands and started the red. This hand moved at approximately three times the speed of the others and thus, due to the spacing of the cross beams, would reach the position of the black hand at the bomb release point. When the black and red hands coincided, an electrical circuit was completed which energised the bomb release mechanism. The accuracy of this system was of the order of 100 yd at a range of 200 miles.

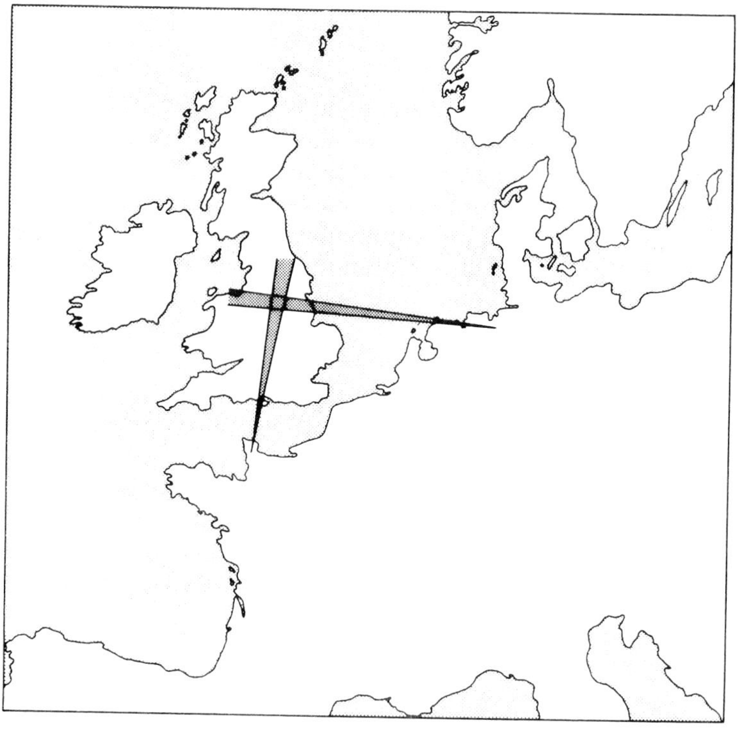

Fig. 4 Knickebein beams aligned on Sheffield

Y-Gerät (Wotan II)

Like its predecessor the X-Gerät, this system was also designed by Dr Hans Plendl. However, the inclusion of a ranging element enabled all navigational information to be transmitted from a single station. The operational frequencies were between 42 MHz and 47 MHz and the ground station transmission consisted of two distinct elements: a directional beam and a ranging signal.

The beam was, in essence, Lorenz-type except that the keying was in the order of 180 per minute and the directional information was conveyed by the relationship of the dots to a synchronising pulse. In the aircraft this was decoded automatically by the receiver which gave a meter presentation to the pilot and also controlled the autopilot which could maintain the aircraft on the beam more accurately than any human hand.

The range element of the transmission radiated a steady tone of 3000 Hz which would be keyed to pass instructions to the aircraft in morse code. When a range check was required, the modulation changed to 300 Hz. On receipt of this lower modulation frequency, the aircraft receiver keyed a transmitter and re-radiated the received modulation on an adjacent frequency. At the ground station the phase difference between transmitted and received modulations were measured on an oscilloscope from which the distance could be calculated. This method could give an accuracy in the order of 100 yd at a range of 250 miles.

The British radio and radar aids

Unlike the Luftwaffe, the RAF had made no attempt before the start of the Second World War to develop radio beams to assist bombing, preferring to rely on traditional navigation methods. By the middle of 1940 it was realised that these methods were not successful as on occasions as few as 25% of bombers actually located their targets, the majority of their bombs falling in open countryside. It was not until the introduction of GEE in 1941 that this deficiency was remedied.

GEE

GEE–the first of the hyperbolic navigational aids–was developed from an idea first suggested by Robert Dippy, a member of Watson-Watts's team at Bawdsey.

This system, in its most basic form, consisted of two transmitters spaced several miles apart, radiating synchronised pulses. A receiver positioned equidistant from the transmitters received pulses from both stations simultaneously. At any other point, the signal from the nearer transmitter was received first and the difference in arrival time corresponded to the difference in path length. It was therefore possible to inscribe on a map of an area a series of lines corresponding to differing signal arrival times. Conversely, if the time difference between the reception of the signal from the two transmitters could be measured, then the position of the receiver could be identified as lying on a particular line. With a third transmitter, two more families of lines could be drawn and by comparing the signal arrival times from all three transmitters, a position fix was obtained.

An immediate problem in the practical realisation of such a system is the identification of the individual transmitters. This could be achieved by frequency separation, but such a method is wasteful in bandwidth. The GEE system therefore used a method of time separation, defining one station as 'master' which transmitted first, followed by one 'slave' at an accurately prescribed time later. The 'master' then radiated a further pulse and lastly a pulse from the second 'slave' whose delay time was different to that of the first 'slave'. With the preset delays entered into the receiver indicator unit the navigator was able to identify and compare the arrival times of the different pulses on a cathode ray tube and ascertain his position with the assistance of a special chart.

The GEE system operated on frequencies between 25 MHz and 80 MHz and was capable of an accuracy of ±2 miles at a range of 350 miles. Although countermeasures made it unusable as a bombing aid after August 1942, it remained in service as a general navigational aid for many years, the last chains only ceasing transmission about 1970.

Oboe

Oboe, the second of the RAF's blind bombing systems, used secondary radar principles, a transponder being carried in the aircraft. This was interrogated by two ground stations, one (code 'Cat') located at Trimmingham in Norfolk and the other (coded 'Mouse') near Walmer in Kent, each of which measured the distance of the aircraft by ordinary secondary radar techniques. Instructions were automatically relayed to the pilot to enable him to fly a circular track, centred on 'Cat' which overflew the target. Operators at the 'Mouse' station assessed the aircraft's ground speed and height and calculated the bomb release point. This was passed to the navigator of the aircraft at the appropriate time.

When countermeasures eventually rendered the Oboe system ineffective on the original working frequency of 220 MHz a later version of the equipment operating on centimetric wavelengths was brought into service. However, the 220 MHz transmissions, although not used, were still radiated and continued to be jammed. By this subterfuge, the enemy never suspected the existence of the centimetric transmissions which remained free of interference throughout the remainder of the war.

The Oboe system had a theoretical accuracy of ±17 yd and was instrumental in showing inaccuracies in alignment between the British Ordnance Survey and pre-war Continental maps.

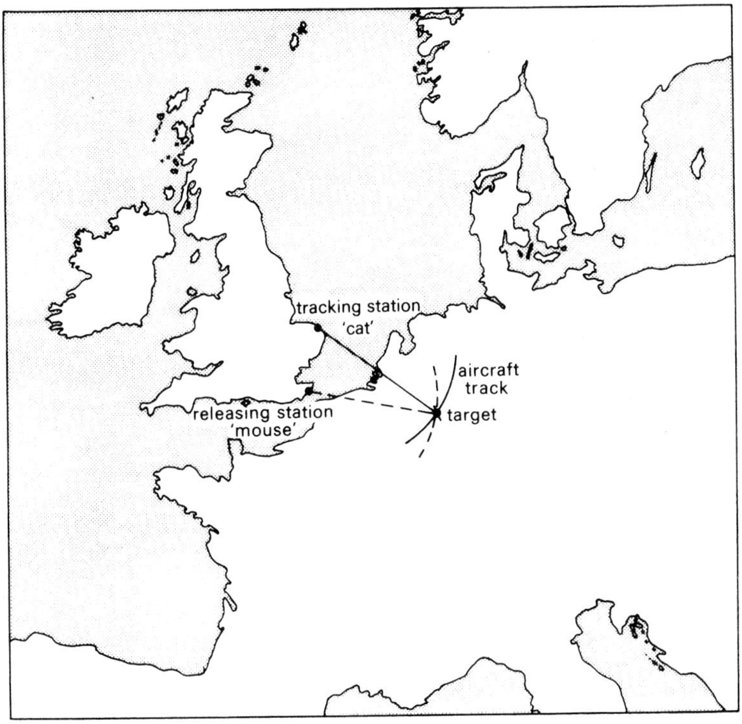

Fig. 5 Oboe

G–H

Oboe had the disadvantage that only one aircraft could be controlled at any one time. This led to the development of G–H which operated on Oboe principles except that the interrogator transmitter was situated in the aircraft and the ground stations were transponders. By arranging that each aircraft had a unique PRF which could be recognised by the receiver circuitry, several aircraft could operate independently using the same ground transponders.

H2S

With the invention of the cavity magnetron by J. T. Randall and A. H. Boot in 1940, the production of a high definition radar sufficiently small to fit into an aircraft was at last possible. The first application envisaged was for use in night fighters but very early in the prototype trials it was noticed that the ground returns from built-up areas were very strong whilst those from water or open countryside were very much weaker. Consequently the question was raised as to whether it would be possible to construct an equipment with a downward-looking aerial, displaying the returns on a PPI such that a passable representation of the ground below would be shown.

Such a set was constructed using a rotating scanner mounted underneath the fuselage of the aircraft. This proved to be of the utmost value and remained in use throughout the remainder of the war. The earlier models operating on 10 cm were replaced in 1943 by an improved version of the equipment using 'X' band (3 cm) with consequent improvement in accuracy. Although not as accurate as the Oboe and G–H system, its value lay in the fact that it was not reliant on ground-based equipment for its operation and consequently could be used on operations far beyond the range of home-based transmitters.

A further development of H2S, designated ASV Mk. III, was used by RAF Coastal Command for the detection of German U-boats travelling on the surface at night. The German naval authorities eventually provided their submarines with receivers capable of picking up these radar transmissions, but the resonant aerials associated with the receivers returned excellent echoes to the aircraft, so their use was to some extent self-defeating.

Section 2
Air traffic management and operational facilities

2.1 The elements of air traffic management

In the early days of aviation, all aircraft could fly on any track at any height which they chose, for their small numbers ensured that there was little risk of collision. This situation still pertains in a few remote areas of the world but in most countries the density of commercial air traffic is such that without strict control, the possibility of collision would be so high as to make the risks involved unacceptable.

This risk was recognised in the early days of commercial aviation and over the intervening years air traffic management procedures gradually improved until the system as we know today reached fruition.

In the management of air traffic, airspace is initially divided into Flight Information Regions (FIR) which are geographical areas under the jurisdiction of a single Air Traffic Control Centre (ATCC). Within an FIR several types of airspace are defined. In the United Kingdom these include danger areas, prohibited areas, two types of controlled airspace and special rules airspace.

Uncontrolled airspace

In uncontrolled airspace in daylight, and provided that certain meteorological criteria are met, an aircraft may fly on any track or at any height which the pilot desires. Under these circumstances he is said to be flying in accordance with Visual Flight Rules (VFR) in Visual Meteorological Conditions (VMC). He is responsible for maintaining separation from other aircraft by visual means and although the Flight Information Service at the associated air traffic control centre will always provide such information as they may have concerning other aircraft movements within the area, this information is not based on radar observation, is purely advisory and the responsibility of maintaining separation lies with the pilot in command.

At night, or when meteorological conditions fail to meet the criteria necessary for VFR, Instrument Flight Rules apply. Under such circumstances aircraft, although in general still responsible for maintaining their own navigation and separation, fly in accordance with the Quadrantal Rule.

Before proceeding further a basic understanding of altimetry is needed. It must first be realised that the altimeters normally used by aircraft are sensitive barometers which have been calibrated in units of altitude, usually feet. As such, these instruments are subject to error due to the variation of atmospheric barometric pressure. To compensate for this, a panel-mounted adjustment device is fitted which also controls a subscale on the face of the instrument. This scale is calibrated in pressure and bears such relationship to the main (altitude) indication that when the main scale indicates zero altitude, the subscale registers the barometric pressure. Thus, should the subscale of the altimeter of an aircraft in flight be set to the barometric pressure at a point on the ground below, then the mainscale will indicate the height of the aircraft above that point.

In practice, various datums for pressure setting are used, the more important being:

(a) Atmospheric pressure at airfield level. This is referred to as the QFE.
(b) Atmospheric pressure measured at a specific point, usually the airfield meteorological office, and corrected to sea level. This is referred to as QNH.
(c) Regional QNH which is a forecast lowest QNH over a specified area.
(d) Standard pressure setting (1013.2 mb), QNE.

The first two settings are normally used when the aircraft is in the vicinity of the airfield and the third (regional QNH) is used when no better pressure setting is available. When flights using regional QNH cross a regional boundary where a different regional QNH is notified, on resetting the altimeter a different height will be indicated. Although this is of little consequence when the flight is under VFR, Instrument Flight Rules require specific heights to be flown and aircraft would therefore suffer the inconvenience of having to climb or descend at each regional boundary. To eliminate this problem, all aircraft flying above 3000 ft in IMC use a standard altimeter setting of 1013.2 mb. When flying on this setting all heights are measured in hundreds of feet and are referred to as Flight Levels, thus 11 000 feet on the standard altimeter setting becomes FL 110.

When flying in IMC or at night, aircraft flying in British airspace are required to comply with the Quadrantal Rule. Above 3000 ft a pilot must maintain a flight level related to his track. This may be summarised as follows:

Magnetic track–000 to 089 degrees	FL 50, FL 70, FL 90, etc.
Magnetic track–090 to 179 degrees	FL 35, FL 55, FL 75, etc.
Magnetic track–180 to 269 degrees	FL 40, FL 60, FL 80, etc.
Magnetic track–270 to 359 degrees	FL 45, FL 65, FL 85, etc.

In order to allow for possible altimeter error at higher altitudes, above FL 245 the rule is modified as follows:

Magnetic track–0° to 179°	FL 250, FL 270, FL 290 and thereafter at intervals of 4000 ft.
Magnetic track–180° to 359°	FL 260, FL 280, FL 310 and higher levels at intervals of 4000 ft.

Controlled airspace

Controlled airspace consists of Control Areas and Control Zones. A Control

Area is a volume of airspace whose geographical area and upper and lower limits are notified and in which air traffic control exists. Where this area takes the form of a corridor between specified points, it is termed an airway.

Airways are normally 10 nm (nautical miles) wide and have their upper and lower altitude limits specified. A confluence of airways in the region of an airport is nominated a terminal control area (TMA) and if several airports are situated within a confined area, a single combined TMA is instituted under a single controlling authority. A typical example of this is within the London TMA below which are six principal airports.

The measure of control exercised over aircraft varies in different areas, this being largely dependent on the type and volume of the traffic normally existent. Much of the controlled airspace over England is, however, notified as subject to one of the rules of air traffic control which requires that Instrument Meteorological Conditions be considered to exist at all times. All aircraft movements are at the discretion, and under the direct control, of the air traffic control authority and before entering such airspace, a flight plan must be filed. Aircraft must fly at the height, track and speed allocated and must not digress without the specific permission of that authority. To ensure that navigation is to the required standard, specified radio equipment must be installed and the pilot must be suitably qualified.

In return, the air traffic control authority ensures that aircraft are separated in accordance with laid down minima and that their journeys are accomplished in an expeditious manner.

In that controlled airspace which is not notified as being permanently subject to instrument flight rules, it is permissible for an aircraft to fly VFR in VMC.

Special rules airspace

While controlled airspace is recognised by ICAO, it is considered unduly restrictive by some elements of the British Aviation community. For this reason, the UK has introduced 'Special Rules Airspace' in certain areas. As the name implies, the rules are tailored to suit the circumstances and full details are published in the 'Rules of the Air' and 'Air Pilot'. Examples of such airspace are: Special Rules Zones associated with some aerodromes, the Cross Channel Special Rules Area and the Special Rules for the Upper Flight Information Regions.

Special VFR flights

A special VFR flight is a flight within a control zone, or within Special Rules airspace where provision is made for such flights, in respect of which ATC has given permission for the flight to be made in accordance with special instructions instead of in accordance with Instrument Flight Rules.

Flight plans

In order to comply with the Instrument Flight Rules, before a pilot flies within controlled airspace, a flight plan must be filed which includes details of aircraft type, radio equipment carried, pilot's name, number of persons on board, survival equipment, route requested, destination and alternate destinations should a diversion be necessary. On receipt, this information is relayed by teleprinter to all air traffic control centres along the route, the destination and alternate airfields. At the air traffic control centres the flight plan information is fed to the Flight Plan Processing System (FPPS) computer which will ensure that the Flight Progress Strips used by the controllers for reference are printed in advance of the aircraft's arrival in the relevant airspace. When a regular schedule is maintained by an airline it is permissable to file a 'Repetitive (stored)' flight plan. This carries all routine details and only requires updating on factors which vary from flight to flight such as number of persons on board and so on. When a repetitive flight plan has been inserted in the FPPS computer, a simple activation message will cause appropriate flight progress strips to be printed throughout the control centre.

On occasions, a pilot originally intending to fly his destination in uncontrolled airspace, will decide to enter controlled airspace. In these circumstances he is permitted to file this flight plan with the air traffic control centre by radio. This is then input to the computer and thereafter handled in the normal way.

Airways

To achieve a safe separation of aircraft in transit and to ensure that they are subject to minimum delay, aircraft are directed along fixed corridors of controlled airspace, known as airways, which connect major airports. Navigation along these routes is by radio beacons (VOR, DME, NDB) and aircraft are required to carry SSR transponders which make them readily indentifiable to the controllers. They are also required to be able to communicate with the air traffic control centre on the appropriate radio channels to enable the controllers to give the instructions for maintaining the required separation standards.

All aircraft flying on airways within the London FIR, which encompasses all UK airspace south of Carlisle, are under continuous radar cover. Separation is therefore possible on the basis of distance and height. Elsewhere, where radar cover is not available, reliance is placed on pilot's positional reports, for example when passing a radio beacon, and separation is achieved on a time basis.

Airfield control

As an aircraft approaches its destination airfield, with permission from its controlling authority, it will leave airways and contact the aerodrome approach control. Under their guidance and depending on the meteorological conditions and the facilities available at the airport, the aircraft will either be vectored by radar, fly an approach pattern utilising specified radio aids or visually make an approach for landing. When aligned for landing, control will be transferred to the airport tower controller in the Visual Control Room (VCR) who will monitor the final approach and landing. When the aircraft has left the runway, instructions for reaching the disembarkation point will be given either by the tower controller or a separate Ground Movements Control (GMC) depending on the size of the airport.

Where sufficient public transport movements exist, additional protection is afforded to the aircraft in the final stages of flight by means of a control zone or special rules zone around the aerodrome, extending from the ground to the base of the TMA.

2.2 Operation of an air traffic control centre

Although no two air traffic control centres are alike, the principles of operation are generally similar and the variations are a function of the area covered and local requirements. A description of the London Air Traffic Control Centre (LATCC) will therefore give a suitable introduction to the operation of air traffic control centres in general.

Airspace in the London FIR is divided into sectors, each of which is controlled from a single radar suite. Some of these sectors are TMAs or part of a TMA and the remainder are en-route which are responsible for airways and the associated upper air routes.

It is a basic principle of control that no aircraft is cleared to enter the airspace of any other sector unless the transfer has been previously coordinated. The controller handing the aircraft over must also comply with any conditions which the receiving controller may lay down.

The TMA sectors are responsible for the control of aircraft arriving at and departing from all the airports within their areas. If an aircraft cannot immediately be accepted by its destination airport, it will be diverted into a 'stack', or holding pattern, which is based on a radio beacon, until such time as it can be accepted. The upper levels of the stacks are controlled by the TMA and the lower levels and the approach by the destination airfields.

The sector teams

Each team is led by a Chief Sector Controller who is responsible for the operations of the sector and liaison with adjacent sectors. He is assisted by several radar controllers who directly control the air traffic within the sector. One of these is a military controller who is responsible for co-ordinating military operational aircraft wishing to use or cross airways or TMAs.

The principle method of control is by radar. Both primary and secondary radar are used, the latter providing electronically-labelled returns indicating aircraft height, and identification. The identification may take the form of either the allocated SSR code or, if the necessary information has been fed into the code-callsign computer, the aircraft callsign. In a similar fashion the aircraft's destination may also be displayed provided that the information has been entered into the appropriate computer.

For information on the aircraft under his control, the controller refers to a series of Flight Progress Strips, which are printed by the FPPS computer on the basis of the information supplied in the flight plan. Initially one strip is prepared for each aircraft on each relevant sector in advance of the time at which it will be required. If operational requirements cause an aircraft to divert from its original flight plan, then the controller need only inform the computer of the change by means of a touch wire display or keyboard and the computer will print corrected flight progress strips to all relevant sectors. In addition to the flight progress strips, the controller also has closed circuit television screens displaying selected additional information which he may need.

Communications

To enable a controller to exercise his function it is essential that he is capable of speaking to all aircraft within his area of responsibility. For this purpose he has the ability to select any two from thirty pre-selected radio channels. In addition, intercommunication is available with any other controller switched to the same channel.

An automatic telephone exchange is also maintained which enables every controller to liase with his colleagues and links him directly with other national and international air traffic control units.

Telecommunications

To provide the facilities and information services required by the air traffic controllers there is necessarily a very comprehensive telecommunications organisation. The day-to-day management of this system is undertaken by the Systems Controller who, working in close co-operation with the ATC watch supervisor, ensures that the telecommunications services meet the requirements of air traffic control at all times.

System control

The system controller operates from a central control room situated adjacent to the controlled airspace operations room. The operations of the Equipment Maintenance Control (EMC) points throughout LATCC and the remote radio and radar stations are co-ordinated from here by means of an extensive private telephone system used solely for technical liaison and control.

The room is equipped to select and monitor the performance of any communication or radar service. The operational status of all sub-systems and facilities is displayed on a tote board which occupies the whole of one side of the control room.

Communications

Controllers communicate with aircraft through a number of remotely-controlled radio stations situated throughout the FIR. These are situated to ensure complete R/T coverage within the region. Multi-carrier techniques are used extensively to provide the coverage required by each radio channel. Each station is connected to LATCC through Post Office Land Lines. In most cases the remote radio stations consist of separate transmitter and receiver sites and contain both VHF and UHF equipment. All facilities are duplicated to ensure that in the case of equipment failure operational service is maintained. Switching from main to stand-by equipment can be achieved by LATCC staff without reference to the staff of the station concerned.

All R/T and some telephone conversations are continuously recorded on magnetic tape recorders and recently this has been extended to include all incoming radar signals. The tapes are usually retained for thirty days before erasure unless further retention is necessary to assist in the investigation of some incident or accident.

Teleprinter services

LATCC is connected by land lines to the Aeronautical Fixed Telecommunications Network (AFTN) and the Meteorological Operations Telecommunications Network Europe (MOTNE). The AFTN connects ATCCs throughout the world and is provided by international agreement for the exchange of aeronautical information such as flight plans etc. The flight plans received at LATCC provide the main input data for the FPPS computer.

The meteorological information required by ATC is received over the MOTNE supplemented by facsimile equipment and distributed via the closed circuit TV system.

Radar services

London Air Traffic Control Centre receives radar information from six radar stations within the UK and one station in Eire. All six UK stations relay both primary and secondary data but the station in Eire, Mount Gabriel, supplies SSR data only. The incoming plot extracted radar is fed to the Processed Radar Display System (PRDS) and from there it feeds the radar displays via display processors.

Distress and diversion, (D & D) Cell

The UHF Emergency Organisation in the London Flight Information Region is controlled from the D & D Cell at LATCC. LATCC controllers have available a country-wide VHF and UHF fixer service operating on the emergency frequencies of 121.5 MHz and 243 MHz, primary and secondary radar and comprehensive direct land lines to the Rescue Co-ordination Centres and other agencies. This enables any incident to be brought swiftly to a satisfactory conclusion.

Plate 7 The C.A.S. control room at the London Air Traffic Control Centre. Each radar suite controls a separate sector of the FIR. *(Photo: Civil Aviation Authority (CAA))*

Plate 8 Part of the IBM 4381 computer at London Air Traffic Control Centre *(Photo: CAA)*

Plate 9 In the Manchester Airport visual control room. On the left of the CCTV information display may be seen the D/F readout. To the right are the wind vane and anemometer dials and the radar distance-from-threshold indicator (DFTI). The buttons control the R/T channel in use or private wire telephone circuit. The controller is updating a flight progress strip. The display on the left is that of the Racal AR18x Airfield Surface Movement Indicator. *(Photo: Racal)*

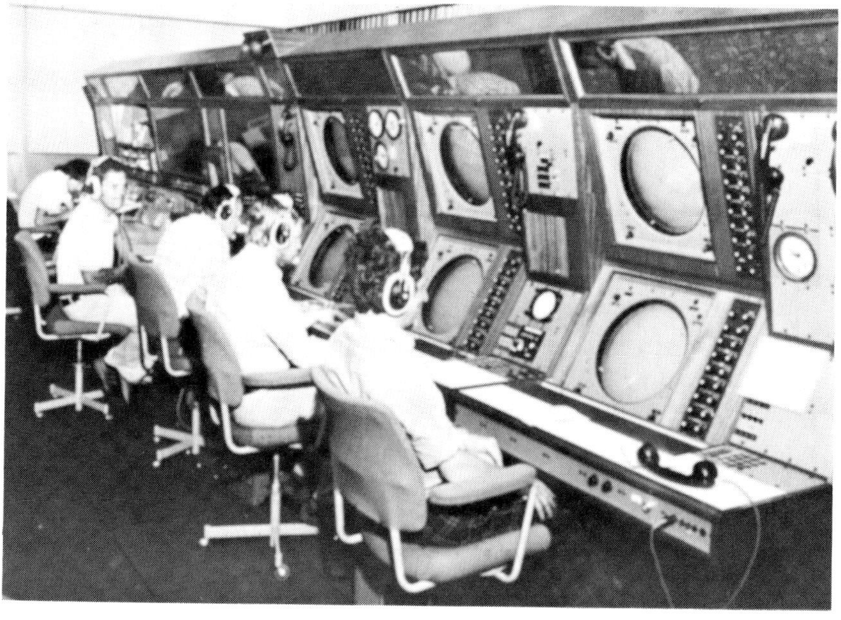

Plate 10 The approach control room at Gatwick Airport. Immediately to the right of each CRT are the display controls. The buttons to the right of each display on the operating bench control telephone circuits while those on the sloping panel are for R/T channel selection. Between the first and second radar displays on the right can be seen the controller and display for the Marconi AD210 VHF direction finder. *(Photo: CAA)*

2.3 Telecommunications organisation on airports

Modern air traffic control procedures are highly dependent on the facilities offered by the telecommunications organisation. It therefore follows that the telecommunications organisation plays a vital role in the day-to-day operation of every airport. The facilities provided are largely dependent on the size of airport, the volume and nature of the traffic and the weather conditions normally experienced. Thus, whilst a relatively small regional airport in Northern Europe might be equipped with ILS, VDF, NDB and radar, a much larger and busier one situated in an area where VMC exists on all but a few days per year, might find a single non-directional beacon (NDB) adequate. Faced with such a wide variation, it is difficult to select an airport which may be described as typical. However, for the purposes of description a medium-sized regional airport such as may be found in many places in the UK or Europe may prove suitable. Due to the prevailing weather conditions, many of these airports possess a wide range of facilities and their organisation is similar to that of their larger international colleagues.

The equipment room

The equipment room may well be described as the nerve centre of the telecommunications organisation for it is here that control and monitoring equipment for the navigational aids and radar and the radio telephony processing, switching and recording equipment are situated. The room is usually manned whenever the airport is operational and staff are required to check at regular intervals the operation of all equipment for which automatic monitoring is not installed. The senior watchkeeper maintains a log of all activities and incidents which may be required for future reference in the event of aircraft incident or accident.

The R/T system

The R/T system at an airport is of primary importance. Discrete channels are normally allocated for such functions as approach control, airfield control, radar, ground movement control and for control of vehicles (normally termed the 'Ground Mobile' frequency). At a busy airport, several of these functions may be allocated two or more channels each.

The air traffic controllers operating these channels may be located in either the visual control room (VCR) or in the approach control room which is usually situated on a lower floor of the control tower. Each control room requires several operating positions and it is desirable that all positions have access to most if not all of the available radio channels. With such an arrangement, a simple paralleling of receiver outputs is not practical for as increasing numbers of positions select a particular channel, so the available audio power at each outlet is decreased. This problem may be further exacerbated if the receiving equipment is situated remotely from the tower. Under such circumstances the receiver output signals are transferred from the receiver station to the tower by telephone lines which have a power level limit of 0.05mW (–13 dBm) which is barely sufficient to satisfactorily drive a single headset.

Similarly, a simple parallel arrangement would not be practical for the microphone circuits for on keying the transmitter, all microphones would become 'live'–a possibly embarrassing situation. Allowance must also be made for the natural differences in level between individual controller's voices.

The problems arising from these factors are solved by the incorporation of line amplifying and switching equipment which is normally situated in the equipment room. The general principle of operation of this equipment may be best described by considering the send and receive paths separately.

Signals from the receivers reach the line equipment via telephone lines at a level of about 0.05 mW. They are then amplified before splitting into individual circuits to the control positions. Further amplification may be included in the individual circuits either to improve isolation or to bring the signals to loudspeaker level.

When a transmitter is selected at the control position, a path is prepared between microphone and transmitter. Amplification may be included both before and after the switching circuitry, the output levels being adjusted such that they are suitable for the telephone circuits connecting the tower to the transmitter site. An additional amplifier with automatic gain control characteristics may be included between microphone and line equipment to compensate for variations in voice level. On depression of the transmit key, the path between microphone and transmitter is energised thus enabling the controller to modulate the transmitter. The switching action only energises one microphone circuit, thus preventing the radiation of casual conversation by other controllers who may also have that channel selected. A further facility frequently incorporated is 'on channel intercom'. This enables conversation between controllers in different control positions which have the same R/T channel selected, this being achieved by energising the microphone circuits and switching their output directly to the input of the receiver line amplifiers.

The function of air traffic control requires frequent use of telephone circuits. For the convenience of controllers, keyboards offering a selection of private circuits and exchange lines are incorporated in each control position. The ear pieces of the controllers headset are individually wired, one being connected to the R/T system and the other to the telephone circuits. The same microphone is used for both functions.

The technique of remote transmitting and receiver siting is in common use on airports. Both sites are situated, as far as possible, distant from the main terminal complex to which they are connected by telephone lines. Preferably the sites are separated by at least a mile to minimise any cross-modulation effect caused by closely adjacent channels. The receiving site should be located in an area of minimum noise (industrial, ignition etc.) to ensure the best possible receiving conditions. On each site, all equipment is duplicated with separate lines connected to each equipment. The telephone lines also are frequently duplicated, reaching the tower by different paths. Further sets of equipment covering the most important radio channels are installed in the tower for use when power failures, line failures or similar circumstances render one or other of the remote stations inoperative.

Navigational aids

The approach aid on a small regional airport would most probably be an ILS located on the main instrument runway, serving the direction corresponding to the prevailing wind. Associated with this would be the standard 75 MHz middle and outer marker beacons. In addition, it is now becoming common practice to install a DME operating in conjunction with the ILS to provide a constant 'distance to run' readout. The categorisation of such equipment would be dependent on many factors, geographical and otherwise, but in general Category 1 or Category 2 would be expected.

Where, for geographical or other reasons, the installation of an ILS system is not possible, approach azimuth (but not elevation) guidance could be provided by a low power (fifty watt) terminal VOR (TVOR) located as near as practicable to the intersection of the runways in the case of a multiple runway airport, or near the midpoint where only one runway is in use.

Telephone lines connected to the individual sites carry signals from the equipment monitors to a central control unit in the equipment room and enable a continuous system integrity check to be maintained by station staff.

To assist aircraft approaching the airport, one or more non-directional beacons (NDB) may be installed adjacent to, but rarely on, the airport. Typically such a beacon may be co-sited with the ILS outer marker. The installations frequently incorporate dual transmitters with an automatic monitor unit which will initiate a changeover from the operational to stand-by transmitter if a decrease in radiated output power or a modulation failure is detected. The signal is also monitored by equipment room staff using a standard communications or specialist beacon monitor receiver. Various means are used to detect which transmitter is in service, the most common being a slight difference in the coding characteristics.

An automatic VHF cathode ray direction finder (CRDF) is also frequently sited on the airport, which, although capable of acting as a navigational aid, is more commonly used to assist in the identification of radar returns. To facilitate both functions, many autonomous radar units can display a CRDF bearing in the interscan period. Alternatively, small (5 in) CRDF displays may be installed at appropriate air traffic control positions.

Radar

The radar installations depend largely upon local requirements. Where surveillance is required over a considerable area, L-Band equipment may be installed, but S-Band or X-Band are more common. The former will give satisfactory surveillance to a range of forty or sixty miles whilst the latter will give highly accurate positional information at shorter ranges. Either of these may be used for a Surveillance Radar Approach (SRA) for which the controller gives directional instructions derived from the information displayed on a PPI radar display to enable an aircraft to make an approach to the runway. Whilst on SRA, the pilot is instructed to descend at specific sink rates in accordance with the progress of the approach as indicated by the radar positional information. X-Band radars are capable of providing guidance to a quarter of a mile from touch-down but the lower accuracy of the S-Band equipment limits their use to minimum ranges in excess of two miles.

It is unusual for an SSR Interrogator to be installed at an airport but the recent increased availability of plot-extracted radar has made the provision of SSR displays a viable proposition. Such signals may be distributed via normal telephone lines to a remote site where two alternative display systems are possible. Most economically a normal SSR display centred on the SSR station may be provided. Alternatively, a display re-centring system could be installed. On receiving the incoming positional information, this equipment transforms the co-ordinates such that the resulting picture may be laid over the airport primary radar display. When the radar station providing the plot extracted video is located at a considerable distance, loss of information at lower flight levels in the local airport area may be expected but in spite of this disadvantage, sufficient information is usually available to facilitate aircraft identification on the primary radar display.

It is possible that, in the future, the availability of plot extracted primary radar used in a similar way will lead to the demise of all but short range X-band radar equipments on airports, reliance for longer range radar information being placed in displays fed from plot extracted signals supplied by major radar stations.

Telephone services

Most regional airports are of sufficient size to justify the installation of a private telephone exchange. In addition, each air traffic control position is equipped with a keyboard which gives access to several extensions from the airport exchange, private wires to nearby airports and the associated control centre and intercommunication lines with other air traffic keyboards.

At some installations the telephone system is effectively divided into two sections–an automatic system catering for the needs of the administrative departments whilst operational requirements are satisfied by a small manual board. This facilitates urgent operational calls gaining priority over those of less immediate importance.

Teleprinter services

Teleprinter services form an important element of airport communications systems for it is by this method that the flight plans for incoming flights are received and those for departing flights distributed. This is achieved by a machine coupled to the Aeronautical Fixed Telecommunications Network (AFTN).

Signals to be distributed by the AFTN are first routed by a direct line to a national switching centre, which in the case of the UK is located at London's Heathrow Airport. Here, a computer examines the incoming signal, and arranges appropriate routing to the ultimate destinations, either directly, if the intended recipients are connected to the centre, or via further switching centres when appropriate. Should the required line be busy, the incoming signal is held in the computer store until transmission can be achieved. All signals passing through the centre are recorded on magnetic tapes which are kept for a minimum of one month before erasure.

The use of automated switching techniques makes it necessary for all incoming signals to adhere strictly to a predetermined format. To help achieve this, AFTN signals are normally sent by autosender controlled by a punched paper tape which has previously been cut on a spare teleprinter within the section.

In addition to the AFTN, increasing use is being made of direct teleprinter links between the Flight Plan Processing System (FPPS) computer at the air traffic control centre and associated airports. Such machines are usually situated within the air traffic control rooms and consequently many of the delays inherent in the manual dispatch of signals between different locations (i.e. teleprinter room to air traffic control) are obviated.

A further teleprinter is allocated to the Meteorological Office Telecommunications Network Europe (MOTNE). This circuit carries weather reports from aerodromes and other meteorological offices throughout Europe. At fixed times, each station attached to the circuit transmits in sequence, thus building up the complete broadcast. As with AFTN signals, those destined for the MOTNE are normally pre-prepared on punched tape.

Within the teleprinter section, additional machines are installed to cater for the Telex tape cutting and training functions.

Communication between the teleprinter section and air traffic control, meteorological office and other essential services is maintained by a pneumatic tube dispatch system except where this method is rendered impractical due to distance or other reasons. Under such circumstances a further teleprinter link is installed between that section and the teleprinter room.

Section 3

Radio telephony communication

3.1 Radio telephony

The most basic aid to aeronautical safety is probably the radio communication between the aircraft and controller.

In the early days of radio communication this was achieved by the use of morse code with the aircraft carrying a specialist wireless operator. The development of VHF radio telephony for short range working and single sideband techniques for longer range has led to the specialist airborne wireless operator becoming redundant as the pilot can now speak directly to the controller.

Modulation

Two methods of impressing speech upon (or modulating) radio frequency signals are used within the civil aviation environment–varying the amplitude in sympathy with the speech waveform (amplitude modulation) and slightly varying the frequency of the signal (frequency modulation).

Amplitude modulation is used on the VHF and UHF aero mobile bands. Formerly this was also for high frequency communications, but has now been superseded by single sideband (SSB), a modified form of amplitude modulation which gives an effective power gain of ten, requires less bandwidth and does not suffer from a type of distortion known as 'phase distortion' which is often prevalent on long distance circuits using amplitude modulation.

Frequency modulation is used on the UHF ground mobile band for airport domestic communications (e.g. Fire Service, ground movements control, marshallers etc.).

Amplitude modulation

An unmodulated radio frequency signal consists of a carrier wave only, occupying an infinitely narrow bandwidth within the radio spectrum. As such it can supply no information, but if that carrier is switched on and off in some predetermined code (such as morse code), the communication can be made. This system is known as continuous wave or more commonly CW and is designated by international agreement A1A.

If, instead of keying the carrier, an audio tone is mixed with the radio signal then it can be proved mathematically that the resultant will be three radio signals–the original carrier wave, a much weaker signal at a frequency of the sum of the radio frequency signal and the audio tone frequency and a third signal at the difference between the two frequencies. For example, if a radio signal of frequency 1000 kHz is modulated with an audio tone of 1 kHz the resultant outputs will be: 1000 kHz, 1001 kHz and 999 kHz. The 1000 kHz is known as the carrier wave and the 999 kHz and 1001 kHz signals are called sidebands. The process of mixing a radio frequency signal with an audio frequency signal is called modulation. Two radio frequency signals may be similarly combined but the process is then known as heterodyning.

If a radio carrier is modulated by a single tone the resultant signal is known as Modulated Carrier Wave or MCW and designated A2A. Within the aviation context this is used for non-directional beacons (NDBs) and ILS marker beacons.

When analysed, human speech is seen to consist of combinations of frequencies of varying amplitude, phase and waveform between about 100 Hz and 5000 Hz although for satisfactory reproduction only those frequencies between 300 Hz and 3000 Hz are necessary. If a radio frequency signal is modulated by a band of signals between say, 300 Hz and 3000 Hz, then the resultant will not be, as in the case of A2A, just three frequencies, but one carrier frequency and two bands of frequencies, one below and one above the central carrier. With a maximum modulating frequency of 3 kHz the overall width of the transmission, known as the bandwidth, will be 6 kHz. This form of transmission is known as Amplitude Modulation (AM) and is designated A3E. All R/T communication on the VHF and UHF aeromobile bands uses this mode of transmission.

Single sideband

After a moment's thought it will be realised that in an AM transmission each sideband is a mirror image of the other, and consequently each contains exactly the same information. Furthermore, the carrier wave contributes no information whatsoever.

If only one of the sidebands could be transmitted instead of two sidebands and a carrier the resultant would be a considerable economy in terms of both power and frequency bandwidth requirements.

This is the form of transmission known as single sideband (SSB) and is designated J3E. Almost all commercial R/T communications within the high frequency waveband now use this mode of transmission.

As mentioned briefly before, SSB does not suffer from phase distortion. This is the effect, common on the high frequency wavebands, in which a received AM signal becomes 'nasal' together with considerable fluctuations in signal strength. The cause of this effect lies with the mode of propagation of high frequency signals. It is well known that high frequency signals travel round the world in a series of reflections between the ionised layers above the earth and the earth itself. The effectiveness of these layers depends on many factors not the least of which is the level of solar activity. The degree of ionisation and height of these layers will affect how efficiently, and which

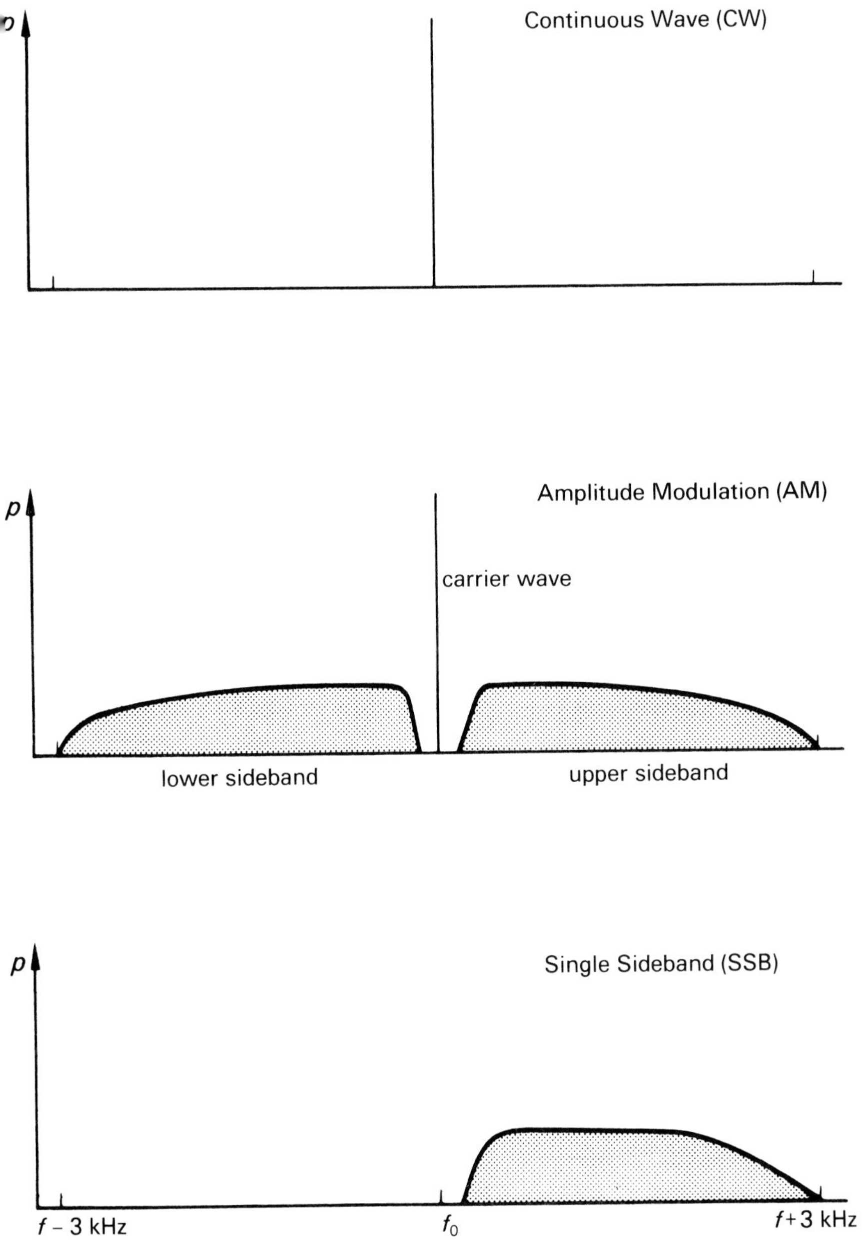

Fig. 6 Bandwidth requirements

frequencies will be reflected. Under certain conditions the two sidebands of an AM transmission are not reflected with equal efficiency and one sideband may be propagated along a slightly longer path. In consequence, the two sidebands arrive at the receiver in differing phase causing partial, or sometimes complete, cancellation in the detection circuits. With only one sideband this effect cannot occur.

Although only the sideband frequencies need be transmitted, the carrier wave has to be re-inserted within the receiver to resolve incoming signals. Immediately prior to, or within, the demodulator stage, a locally-generated signal is mixed with the incoming SSB signal to substitute for the missing carrier wave. The stage generating this signal is known as the Beat Frequency Oscillator (BFO) and should the output of this stage not be of the correct frequency within very fine limits, the resolved speech will seem unnaturally low or high-pitched as the case may be. A panel control on the receiver marked either BFO or 'Clarifier' is usually fitted to SSB receivers to correct this effect. An alternative method used on some commercial communication circuits is to transmit the carrier wave at a considerably reduced level. In the receiver this is used to synchronise the BFO.

Another adaptation of the single sideband principle is the mode known as Independent Side Band (ISB) in which effectively two single sideband transmissions, each conveying different information, are operated on either side of a common carrier frequency.

Speech processing

If a speech waveform is examined on a cathode ray oscilloscope one of the most noticeable features is the wide variation of levels in the different parts of the waveform. Translated into terms of modulation of a radio signal it means that whilst some of the speech waveforms fully modulate the radio frequency carrier, others make so little impression that the received signal has to be of considerable strength for them to be demodulated satisfactorily in the receiver. Experience has shown that the variation between the different parts of the waveform can be reduced considerably without adversely affecting the intelligibility apart from a slight change in 'timbre', but with the overwhelming advantage of an considerable increase in the 'loudness' of the speech. In a Radio Telephony transmitter this results in a much higher average modulation depth with consequent advantages to the receiving station particularly in conditions of low signal strength or high ambient noise level. This action of modifying speech waveforms is known as 'speech processing'.

Three types of speech processing are in common use–the AF clipper, the compressor and the RF clipper, the first being the simplest and the last the most complicated.

The operation of the AF clipper is, as the name suggests, that any waveforms above a predetermined level are electronically 'clipped' with the result that the ratio of the strongest to the weakest waveforms is considerably reduced. This is not achieved without penalty as the clipping action effectively 'squares' the clipped waveforms. An analysis shows that a square waveform consists of a fundamental frequency and all harmonics of that frequency, thus the clipping action will produce considerable harmonic content within the audio spectrum with consequent distortion of the speech. The amount of distortion permissible will set the limit on the level of clipping possible. Nevertheless for economy, this method has much to commend it and is frequently used in less expensive equipment.

In an audio compression circuit a somewhat similar effect is achieved by arranging that the gain of the amplifier rapidly reduces when the input signal approaches a preset level. By this method the mean signal strength compared with peak can be considerably increased. A disadvantage of the compression is that in speech waveforms peak levels are frequently followed by weaker elements that are important to intelligibility. The time constants of the gain control circuitry are such that these weaker elements of speech may also be subjected to the same gain level as the preceding speech peak and

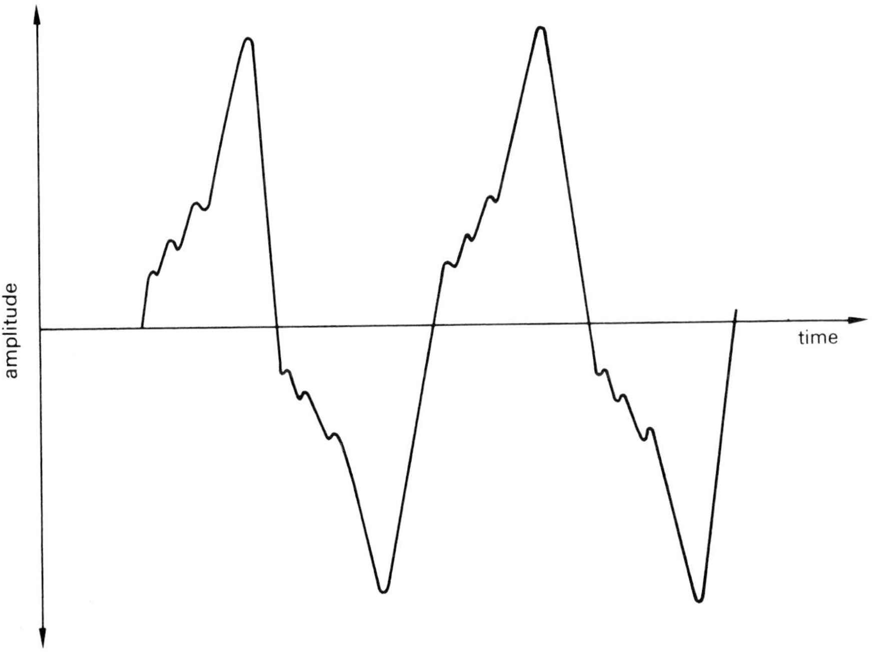

Fig. 7 Typical speech waveform

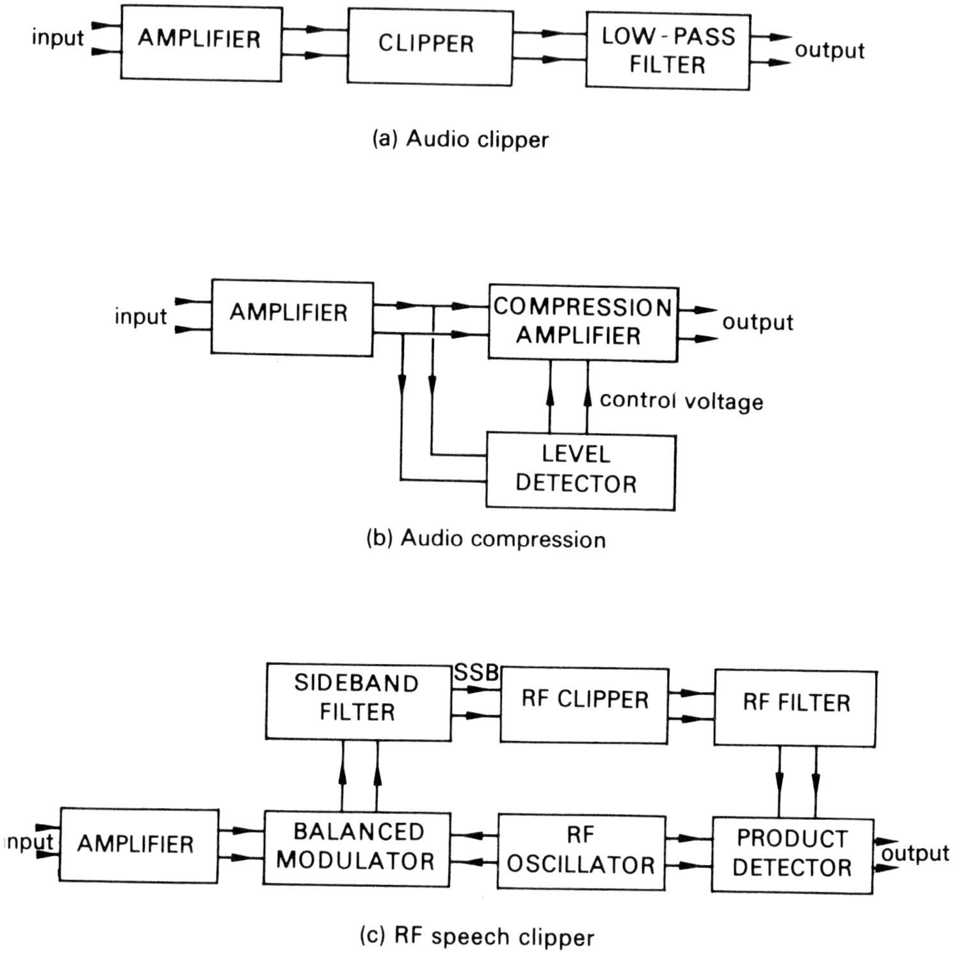

Fig. 8 Speech processing circuits

consequently considerably attenuated thus causing a reduction in intelligibility. Despite this disadvantage it has proved effective and many of the older VHF transmitters are fitted with this type of processing.

In recent years RF speech clipping has gained in popularity, as this method permits a higher degree of clipping coupled with less speech distortion than either of the previous methods. In an RF speech clipper the incoming audio signal is first amplified and then converted to an SSB signal at some convenient frequency. This signal is then clipped in a similar way to an audio clipper with an important exception. Whereas in an audio clipper when a signal of, for example, 500 Hz is clipped it will produce harmonics of 1000 Hz, 1500 Hz, 2000 Hz, 2500 Hz and 3000 Hz within the audio spectrum, all contributing to distortion; in the RF Clipper, if the SSB carrier frequency is 60 kHz, then a 500 Hz audio signal will produce a sideband of 60 500 Hz and the harmonics produced by the clipping action will be 121 kHz, 181.5 kHz and so on which can be removed by a very simple filter circuit. Subsequent to this filter the signal may be either heterodyned to the final operating frequency–in the case of an SSB transmitter–or, alternatively, restored to the audio spectrum for driving any other type of transmitting equipment.

Lincompex

A highly efficient form of speech processing known as 'Lincompex' is used on many high frequency point-to-point circuits. This name is derived from the words LINked COMpressor and EXpander and thus in a single word describes the principles of the system.

Whereas the speech processing systems so far described seek only to increase the average level of modulation at the transmitter, the Lincompex system also expands the compressed signal received, thus restoring the dynamic range originally presented to the transmitter.

The heart of the transmission equipment lies in a rapid acting compressor which will maintain a constant output–even at syllabic rate–to drive the transmitter. At the receiver, the incoming signal is first brought to a constant level by a fast-acting, fading regulator before re-imposing the dynamic range originally present at the transmitter input.

To link the action of the send path compressor and the compensating receive path expander, a control tone is added to the speech channel on the radio circuit, normally at a level 5 dB below the constant level speech. To permit this the speech channel is arranged to have an upper cut-off frequency of 2700 Hz and the Control Tone Frequency is varied between 2840 Hz and 2960 Hz, rising by 2 Hz for every 1 dB fall in instantaneous speech input level. Thus a 60 dB dynamic range of input can be handled without operator intervention and allowing the overall circuit from Lincompex input to Lincompex output to be used in the same way as a cable circuit.

VOX (voice operated transmission)

In a heavy workload operational environment it would obviously be of considerable advantage to relieve the operator of the necessity of having to physically depress a switch on every occasion that he wished to transmit. This end may be achieved by use of a VOX circuit in which an audio ouput is taken from an early stage of the audio amplifier. It is then amplified and fed to a rectifier circuit, the output of which is used to operate the transmit/receive relay. In more modern equipment, an output from the audio amplifier operates an electronic trigger circuit whose change in output level operates solid state switching circuits throughout the transmitter thus obviating the need for electro-mechanical relays.

Two controls are necessary on a VOX circuit–threshold and delay, or 'hang' time. The former is necessitated by the fact that any open microphone will pick up distant sounds. A circuit must therefore be designed so that these relatively low-level sounds do not trigger the VOX circuit. The control that sets the minimum level that will operate the VOX circuit is called the 'VOX threshold' or 'VOX sensitivity'.

Once operated it would obviously be undesirable for the transmitter to switch off at the end of each syllable or momentary break in speech. To maintain transmission during these short breaks a circuit is incorporated that delays the switch-off after the last syllable of speech. This delay can be adjusted from about 0.1 s to 3 s by a control called 'VOX delay', the period being normally set to about $\frac{2}{3}$ s.

Equipment design

Although ground and airborne transmitters operate on the same frequencies and use the same modes of operation, due to the difference in environment and operational requirement there are considerable differences in design. Ground equipments are normally of more simple design than the corresponding airborne equipments, consequently it would be advantageous to consider these first.

VHF aeromobile band ground transmitters

The basic requirements for an aeromobile band ground transmitter are that the equipment is capable of radiating an amplitude modulated signal of high stability containing low spurious content and that it is capable of being controlled remotely via telephone lines.

Frequency control is by quartz crystal. A number of crystalline substances have the ability to transform mechanical strain into an electrical charge and vice versa. This property is known as the piezoelectric effect.

If a small plate or bar of such material is placed between two conducting electrodes it will be mechanically strained when the electrodes are connected

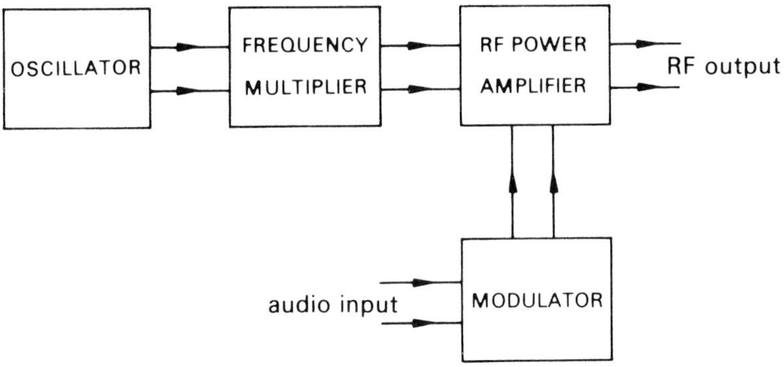

(a) Basic AM transmitter

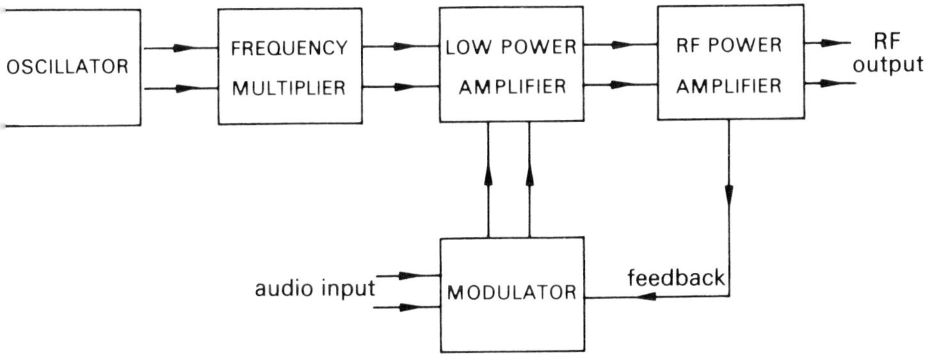

(b) Low level modulation AM transmitter

Fig. 9 Basic types of AM transmitter

to a voltage. Conversely, if the crystal is compressed between two electrodes a voltage will develop across those electrodes, thus piezoelectric crystals can be used to transform electrical to mechanical energy and vice versa. This effect is frequently used in inexpensive microphones, gramophone pick-ups, and in some headphones and loudspeakers. For these purposes crystals of Rochelle salt are used. Crystalline plates also exhibit a mechanical resonance of frequencies ranging from a few thousand to many millions of cycles, the frequency depending on the material of the crystal, the angle at which the plate was cut from the crystal and the physical dimensions of the plate. Due to the piezoelectric effect the plates also exhibit an electrical resonance and act as a very accurate, and highly efficient, tuned circuit. Such crystals are used in radio equipment in high-stability oscillator circuits and in highly selective filter circuits.

In ground transmitters a quartz crystal is used as the frequency determining element, controlling the oscillator circuit. Until recently it has not been possible to manufacture a crystal that will resonate in the 110 MHz to 136 MHz aeromobile band and even now these crystals are fragile and expensive. A crystal on a submultiple of the required frequency is therefore often used, followed by circuits designed to multiply the oscillator frequency to the final operating frequency. In early days, crystal frequencies of 5 MHz to 7 MHz were used requiring multiplication factors of 18 or 24 times, but more recently, improved techniques in cutting and mounting have produced much higher frequencies, requiring far less frequency multiplication. The crystal multiplier chain is followed by amplification stages which generate the necessary output power.

The circuits that impress the speech on the transmitted RF carrier wave are known collectively as the modulator. Essentially consisting of a specially-designed audio amplifier, speech processing, VOX and audio filtering circuits are also included. The output power required depends upon the point in the RF power amplifier circuitry at which the modulation is applied. In older transmitters using valve technology the frequency multiplier chain gave sufficient driving power to drive the Power Amplifier (PA) valves directly. Consequently it was only possible to modulate the output valves directly, the normal method being to insert a winding of the modulator output transformer in the power feed to the PA valves. The effect of this was that the output voltage from the modulator added to, or substracted from, the voltage applied to the PA thus varying its power output in sympathy with the incoming speech waveform.

With the advent of transistor equipments this technique was again applied but recently an alternative method has come into use.

Transistors are not at present capable of the same power gain as valves and furthermore, in other than power circuits, commonly operate at far lower (about 1/20) power levels, thus, whereas a valve frequency multiplier chain would provide perhaps 1 watt to 2 watts power for driving the PA valve to an output of 50 watts or more, the transistor multiplier chain will only give an output of the order of 50 mW to 100 mW. Power transistors will normally only give a power gain of the order of 20 and consequently two or more stages may be required to build up to the necessary output power. This makes possible the application of modulation in an earlier, low power stage of the equipment, with consequent savings in modulation power necessary. Where this method is used it is normal to feed a small amount of the final PA output power as negative feed back to the modulated stage to correct for any inadvertent distortion.

Frequency modulation transmitters

Within the aviation context, frequency modulation transmissions are used for ground mobile (e.g. fire service, marshallers, maintenance workers, etc.) equipment on the UHF band (440 MHz to 460 MHz).

Frequency modulation is, as the name implies, a slight variation of the transmitted frequency in sympathy with the speech waveform. The RF power output of the transmitter remains constant at all times.

The frequency of a crystal oscillator can be varied very slightly by placing a small capacitor in parallel with the crystal. Under certain conditions a valve or semi-conductor can be made to act as such a capacitor whose value varies in accordance with applied drive voltage. These two effects are used together in a circuit known as a reactance modulator to cause a slight frequency modulation of the oscillator, the output of which is then multiplied typically 24 or 36 times to reach final operating frequency.

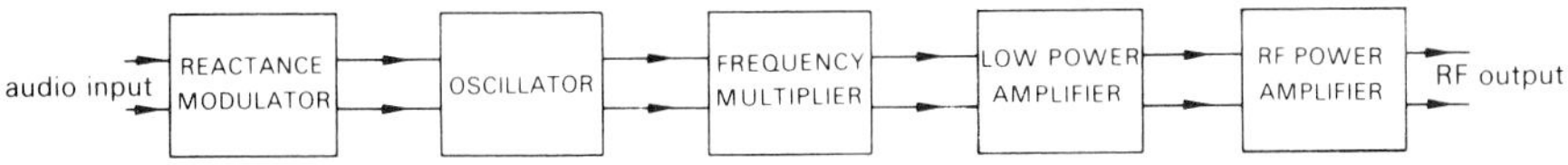

Fig. 10 FM transmitter block diagram

After frequency multiplication the signal is fed to the Power Amplifier stages and from thence to the aerial. In FM service the Power Amplifier devices, whether valves or semiconductors, do not need to be as high a rating as for an equivalent power AM transmitter for, whilst in AM service the output devices have to be capable of handling the power necessary for both carrier and sidebands, in FM service the output devices operate at a constant level and consequently may be operated nearer to their maximum ratings.

Phase modulation (PM)

It is often more convenient to modulate the phase of a transmission rather than the frequency. This is known as phase modulation (PM) and is achieved by the addition of a modulator located between the oscillator and the succeeding stage.

Although there are slight variations between the characteristics of FM and PM transmissions, they are, to all intents and purposes, fully compatible.

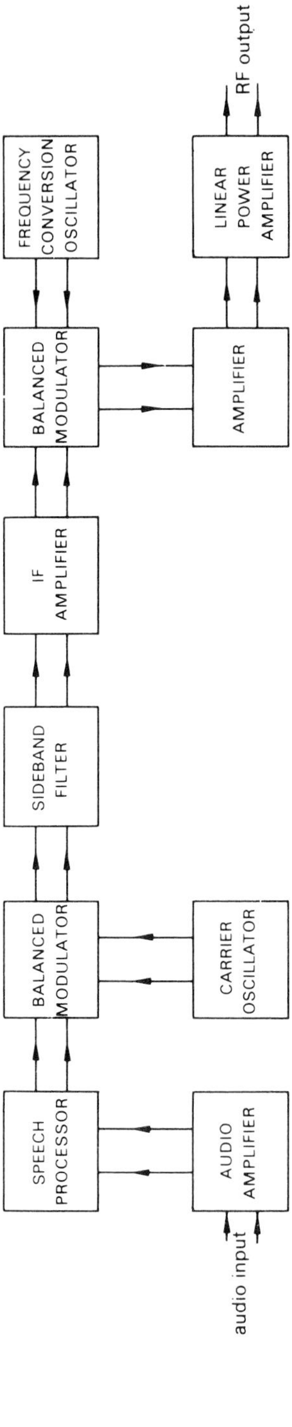

Fig. 11 Block diagram of typical single sideband transmitter

Single sideband transmitters

Single sideband (SSB) transmissions are used in the HF aeromobile band and for point-to-point services. As described previously, an SSB transmission is the residue of an AM transmission after the carrier and one sideband have been removed. The techniques for generating SSB are somewhat complicated, there being three distinct methods: phasing, filtering and the so-called 'third method'. In many ways the filtering technique is both the simplest and most commonly used and this method is described here.

A carrier is generated at some convenient frequency by a crystal oscillator. This and the output from a low power audio amplifier are fed to a circuit known as a balanced modulator, the action of which permits the audio signal to modulate the generated carrier to produce the normal AM signal of carrier plus upper and lower sidebands and then balances out the carrier leaving only the two sidebands in the output circuit. This signal is then applied to a highly selective mechanical or crystal band pass filter which effectively removes one of the sidebands, leaving a single sideband suppressed carrier signal. Final output frequency is attained by heterodyne action, followed by amplification to the required output power level.

Receiver principles

It has been explained earlier in this chapter that an amplitude modulated signal comprises a carrier wave plus upper and lower sidebands. This is a description comparing power output with frequency. In some cases, however, it is more convenient to compare power with time. In this case the unmodulated carrier shows a constant level and when modulation is applied the power level varies in sympathy with the speech waveform.

The most elementary form of receiver consists only of an aerial connected to a rectifier and a pair of headphones, the action of the rectifier being to extract the audio component of the modulation envelope. Due to the lack of sensitivity and tuning arrangements such simple equipment is unsuitable for anything other than monitoring very strong local signals. The fitting of a simple tuned circuit will convert this simple receiver into a present-day version of a crystal set and, with a suitable aerial and earth, this is capable of receiving strong MF signals such as medium wave broadcast stations. To improve this basic circuit further, amplification may be added both between aerial and rectifier (normally known as 'detector') to improve the ability to receive weak signals, and after the detector to increase audio power output even to the stage where a loudspeaker may be used. This arrangement is known as a TRF (Tuned Radio Frequency) circuit and was in common use between 1920 and 1939.

From an operational point of view the TRF circuit had severe limitations, for, as the tuned frequency was raised, it became increasingly difficult to achieve sufficient amplification without instability and the increasing number of stations active created a need for increased sharpness of tuning, i.e.

selectivity. This called for more tuned circuits, which had to be 'ganged' together with consequent difficulties in maintaining accurate tracking over the tuning range of the receiver. These factors led to the introduction of the superheterodyne receiver and TRF equipment faded into obsolescence. The principle of the superheterodyne receiver (normally referred to as a 'superhet') is that after a small degree of amplification the incoming signal is mixed with a locally produced signal. The consequent heterodyning action produces a third signal at a fixed intermediate frequency (IF) which carries the same information as the incoming signal and which may be amplified as necessary. Due to the IF frequency being predetermined, sufficient fixed tuned circuits, mechanical or crystal filters may be included to achieve the degree of selectivity required for the equipment to perform its purpose.

Difficulties can arise with the superhet receiver and as an example of this consider an incoming signal at a frequency of 10 MHz, which is being mixed with a locally-generated signal of 11 MHz to produce an IF of 1 MHz. Unfortunately an incoming 12 MHz signal mixed with the locally produced 11 MHz signal will also develop and IF of 1 MHz and will, in consequence, interfere with the desired signal. Separation of the wanted and unwanted signals is achieved by tuning the signal frequency stages of the receiver. If the IF is low, at HF and VHF, simple tuned circuits may be inadequate to provide sufficient selectivity. The greater the frequency separation between wanted and unwanted signals (called the 'image' or 'second channel' frequency), the less the difficulties in achieving sufficient selectivity. It is therefore desirable for the IF to be as high as possible, although to facilitate maximum selectivity with minimum complication a low IF is preferable. Thus there are two conflicting requirements. In receivers operating only in the MF and LF bands few problems arise because an IF in the order of 450 kHz to 1.4 MHz will satisfy both requirements but in the HF and VHF spectra the confliction in requirement has resulted in the heterodyning principle being used twice within the same receiver. Firstly from the signal frequency to a high IF to minimise interference from signals on image frequency and then to a low IF to achieve the necessary degree of selectivity. In the extreme case some HF receivers in the late 1950s were produced with three IFs, e.g. 3.5 MHz, 455 kHz and 50 kHz. Such equipments are known as double-superhets or triple-superhets as appropriate.

The development of high frequency crystal and mechanical filters in recent years has led to a reversal of technique and many of the more recent VHF equipments are again of the single superhet configuration.

Incoming signals are not all of similar signal strength and any one signal may vary considerably over a short period of time. Under these circumstances the gain of the receiver is in continuous need of adjustment. To achieve this requirement an Automatic Gain Control (AGC) circuit is incorporated. An additional detector circuit is connected in parallel with that used to recover the audio component of the received signal but, in this case, the circuit constants are chosen such that it will not respond to the audio but to the mean strength of the signal. The output of this circuit is then used as a control voltage to reduce the gain of the earlier stages of the receiver thus tending to minimise the effects of variation of signal strength. More comprehensive receivers often incorporate an additional amplifier to facilitate more effective control. In addition a meter presentation of this voltage may be made to indicate the strength of the incoming signal. AGC may also be generated within and to control the audio stages of the receiver for even more effective operation. It is not uncommon in modern equipment for a 100000:1 variation of incoming signal level to make no perceptible variation in audio output.

For some purposes it is frequently necessary to monitor a quiet channel for long periods. In these circumstances the background hiss, and in some locations ignition or other spurious noise, can cause intense irritation and indeed fatigue to the operator. For this reason most VHF equipments incorporate a muting (sometimes called 'squelch') circuit which disables the audio amplifier of the receiver in the absence of an incoming signal. The level at which the circuit operates can normally be adjusted to suit local conditions.

Frequency modulation (FM) receivers are broadly similar to the AM equivalent with the exception that the detector is replaced by two circuits: the limiter followed by the discriminator. An FM signal is transmitted at constant amplitude but due to the vagaries of propagation between the transmitting and receiving aerials the received signal may vary considerably in signal strength. The receiver is operated in a high gain condition and the design parameters of the last IF stage are such that, as far as possible, all incoming signals cause this stage to saturate. The effect of this is that, regardless of incoming signal strength, this stage (known as the 'limiter') feeds a constant strength signal to the discriminator.

Whereas an AM detector gives an audio output voltage in sympathy with the variation of strength of the incoming signal, the discriminator gives an output voltage which is sympathetic to the variation in frequency of the incoming signal. This voltage is then amplified in the audio stages as in an AM receiver.

VHF AM ground receivers

A VHF AM ground receiver is required to operate on one channel only, unattended, in a building often some distance from the point where the operator (or Air Traffic Control Officer) is located.

These requirements lead to a superhet design in which the local oscillator was crystal controlled. In early equipments a crystal multiplication factor of 12 or 18 was used but latterly the development of harmonic cut crystals has enabled the factor to be reduced to two or three. More recently, however, with the decreasing cost of integrated circuits, a preset frequency synthesiser has replaced the simple oscillator chain in many modern designs of equipment.

The frequency changer stage is followed by a high frequency crystal or mechanical filter to achieve the necessary selectivity. Sufficient IF amplifi-

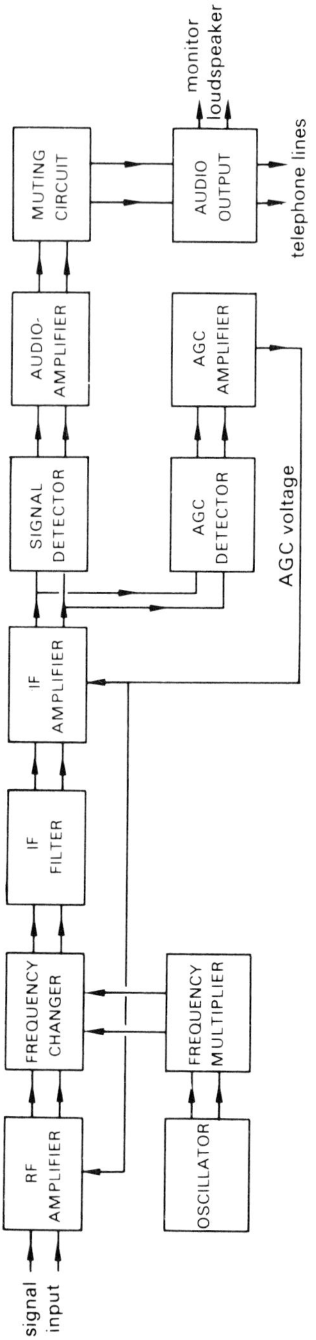

Fig. 12 Block diagram of typical VHF ground receiver

cation stages for the required channel gain complete the signal path. Audio output, amplified AGC and muting stages complete the circuit.

The output stage feeds two separate outputs, one to a local loudspeaker for local monitoring and a second to match telephone lines transferring the signal to the Air Traffic Control or operating position.

Since the advent of transistorised receivers it has been normal to power these from rechargeable batteries which are kept on continuous charge from a mains powered charging unit. This is to safeguard the service against a failure of the incoming mains voltage. Should such a failure occur, the receivers are capable of operating for at least 24 hours from the battery supply.

It is usual to install two receivers or more on each operational frequency so that in the event of receiver failure, service may be maintained on the stand-by equipment.

Receiving stations are frequently situated several miles from both the main air traffic unit and the transmitting stations. The advantages stemming from this arrangement are: that the location may be selected for freedom from electrical interference and the separation of the transmitters and receivers prevents any mutual interference between adjacent channels. The principal disadvantages of the arrangement are: the provision of the necessary building and works services and the reliance on telephone lines for connection between receiver stations and the air traffic control unit. Airfields are notorious for continuing development programmes which involve the use of heavy earthmoving machinery. Occasionally underground telephone cables are inadvertently cut, severing many circuits. To mitigate any such potential disaster, further receivers on the most vital frequencies are fitted in the control tower, but these are rarely used except in an emergency.

UHF FM receivers

Operating in the ground mobile band of 440 MHz to 460 MHz, these receivers are in many ways similar to the VHF AM receivers; the only concessions being the replacement of the AM detector by an FM limiter and discriminator, additional frequency multiplication in the local oscillator stages and slightly redesigned signal frequency and mixer stages to accommodate the more critical nature of the higher frequency involved.

HF Single Sideband reception

On single sideband radio circuits reception may be on either a standard communications receiver, a crystal controlled specialist receiver or the receiving element of a tranceiver.

All communications receivers are built to the superhet principle, but emphasis may be placed on differing features in accordance with the operational requirement.

The main requirements for an SSB receiver are the provision of a method of re-insertion of the suppressed carrier (i.e. a BFO), extreme frequency stability, a bandwidth of about 2.5 kHz with sharp cut-off on either side, and adequate sensitivity.

Recent developments in integrated circuit technology have revolutionised the design of the oscillator stages of many transmitters and receivers. The design of the oscillator is such that it is tuned by a varactor diode. This type of diode is specially designed to operate as a variable capacitor when incorporated in reactance modulator circuits (as described under FM transmitters), thus tuning may be achieved by application of a variable voltage to the varactor diode. Such a circuit is known as a Voltage Controlled Oscillator (VCO).

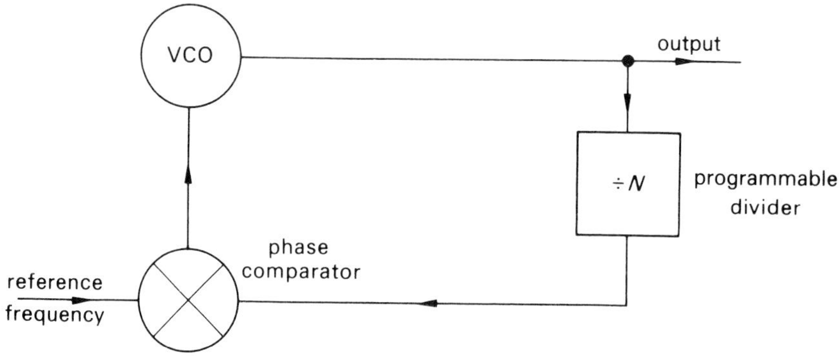

Fig. 13 Basic frequency synthesiser

The output of this circuit is split, one output being to the mixer (in receivers) or amplifier/multiplier stages (in transmitter service). The other output is to a programmable digital frequency divider chain. The output of this chain is compared with that of a crystal controlled reference in a phase comparator circuit, the error voltage being used to control the VCO. The VCO is thus locked to a multiple of the reference oscillator frequency, the multiple being decided by the programming of the digital frequency divider chain. It may seem impractical to tune a receiver by discrete steps but as these steps may be as small as 10 Hz, to all intents and purposes tuning is continuous, with the stability of crystal control on all frequencies. The output frequency of the synthesiser is controlled either by mechanically switching the control circuit inputs, in which case the required frequency is selected by the positions of a number of rotary switches mounted on the receiver front panel (usually seven), or electronically from a digital encoder which effectively operates electronic switching of the control circuits. The frequency selected is indicated on Light Emitting Diode (LED) numericators and the operation of the digital encoder is by a knob which seems to the operator like the tuning knob of yester-year.

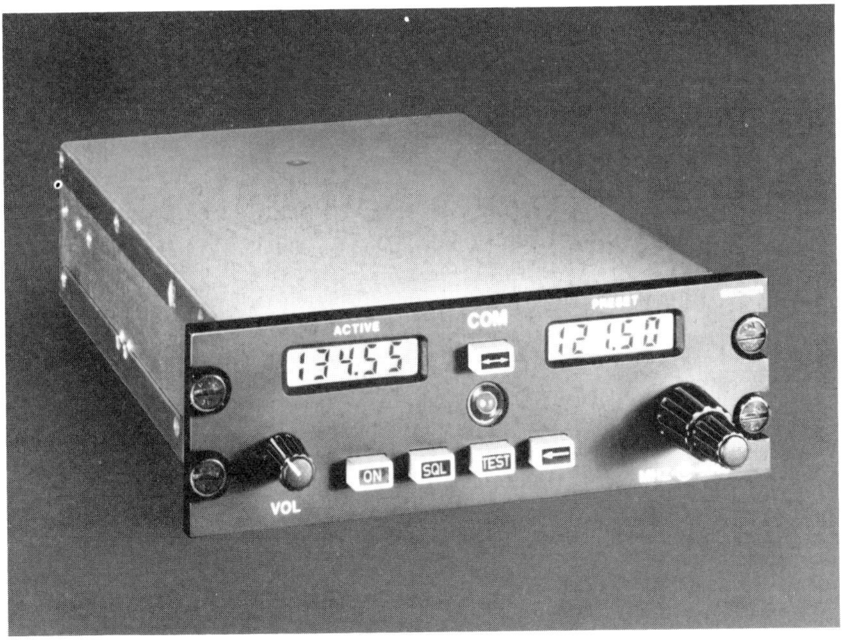

Plate 11 The Becker AR3202 VHF Transceiver for aircraft. This equipment operates on any of 760 channels between 115.0 and 136.975 MHz, one of which may be preselected. The transmitter has an output of 20 W. *(Photo: Becker Flugfunk)*

Not all receivers use synthesisation, however, some obtaining their stability by use of an Automatic Frequency Control (AFC) circuit. This operates from a discriminator circuit fitted in parallel with the demodulator. Should the incoming signal not be in the centre of the passband of the receiver, a voltage proportional to the frequency error is generated which is used to correct the tuning of the local oscillator of the receiver.

The necessary stability was obtained in many older receiver designs by using the double or triple superhet principle, crystal controlling the first conversion oscillator and tuning the first intermediate frequency. As this oscillator tuned only across one frequency band, considerable attention could be paid to the design and these receivers were often extremely stable.

For many purposes the ability to tune across a wide band is unnecessary, only a few predetermined channels being required. To meet this requirement many manufacturers can supply receivers using a crystal controlled local oscillator, a mechanical switch selecting the appropriate crystal for the channel in use. In all other respects these receivers are similar to their tuneable counterparts.

The frequency changer stage is followed by the selective filters. These may

Plate 12 The Collins 51Y-7 airborne Automatic Direction Finder, an ADF equipment operating on the crossed loop principle. The dual frequency selector permits preselection of required channels. *(Photo: Collins)*

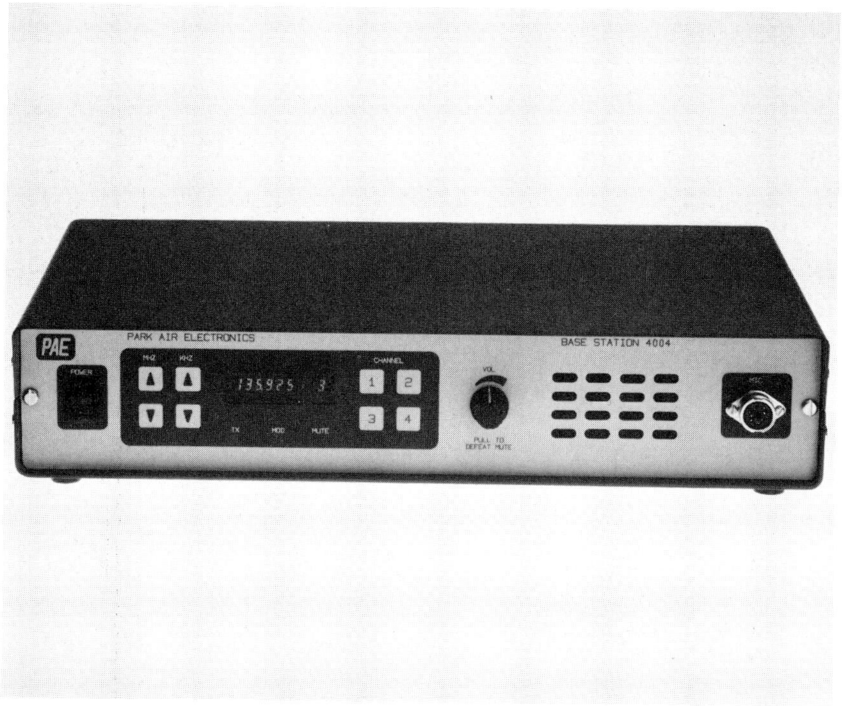

Plate 13 The Park Air Electronics Type 4004. This is a transmitter-receiver for the VHF aeromobile band. Incorporating synthesised frequency control, it is capable of operation on any of four channels which may be selected by the front panel controls. *(Photo: Park Air Electronics Ltd)*

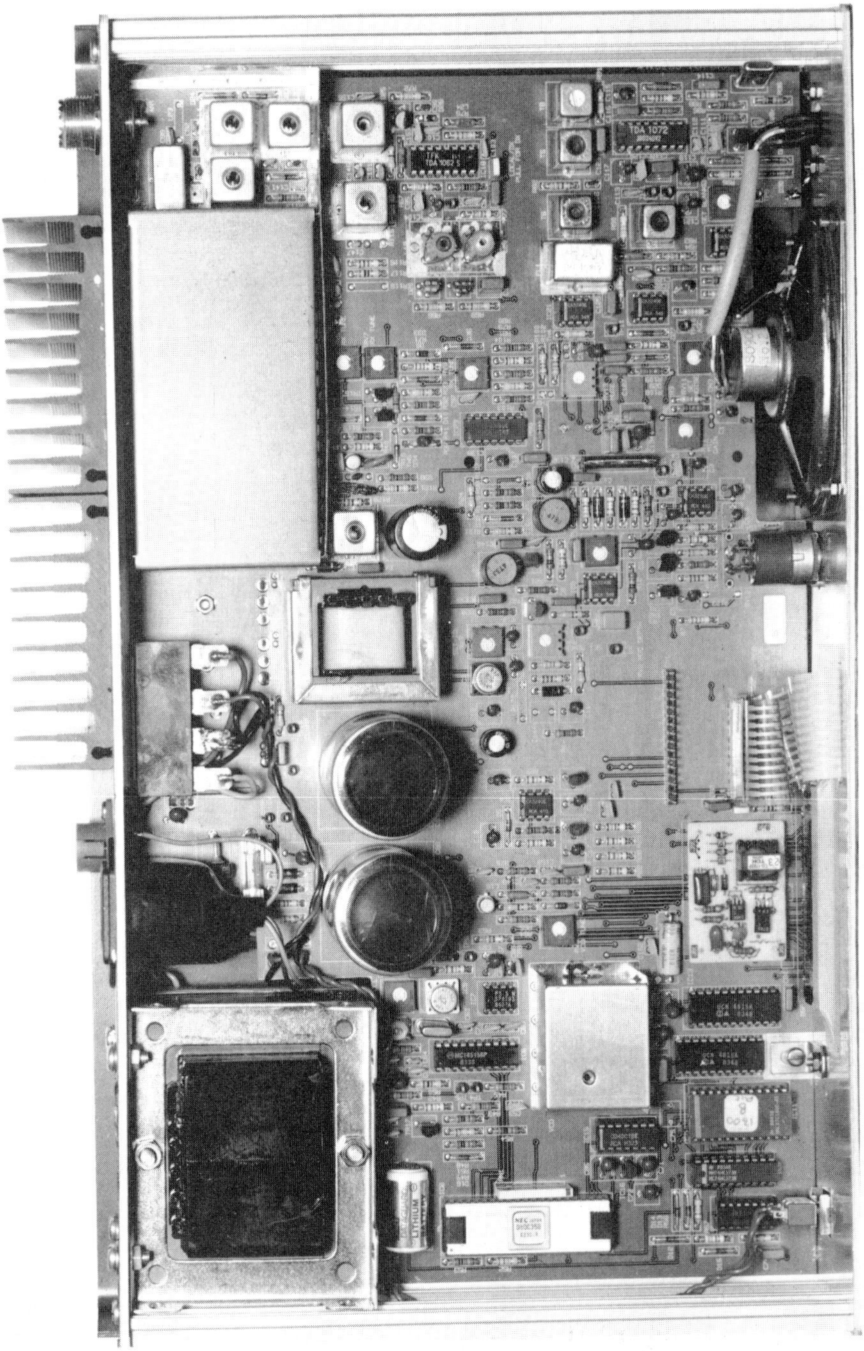

Plate 14 The inside view of the Park Air Type 4004 equipment *(Photo: Park Air Electronics Ltd)*

Plate 15 The Racal Communications RA3701 general purpose modular HF communications receiver which covers 15 kHz to 30 MHz in 10 Hz steps *(Photo: Racal)*

Plate 16 A typical 1 kW transmitter such as is used for point-to-point or air-to-ground service. *(Photo courtesy Racal Avionics)*

Plate 17 Doppler VHF direction finder. *(Photo: Fernau Electronics)*

be either mechanical or crystal resonators designed to pass the band of frequencies necessary to receive the desired signal and severely reduce the strength of any signals outside this band. A separate filter is required for each bandwidth required and two are required for SSB, one for USB and one for LSB. A typical high quality communications receiver may accommodate six or eight filters of different bandwidths for the reception of CW, MCW, SSB and AM. The filters are succeeded by the IF amplification stages and demodulation in a circuit known as a product detector. In this stage the output of the BFO is substituted for the carrier suppressed in the transmitter and the resultant pseudo-AM (i.e. carrier and one sideband) is demodulated as in the AM receiver. Audio amplification and power output stages complete the circuitry.

Tranceivers

There is frequently confusion between the terms transmitter-receivers and tranceivers. In general the former refers to equipments which contain a transmitter and a receiver in the same housing – probably but not necessarily sharing sections of the same power supply. In a tranceiver, on the other hand, certain elements such as oscillator tuning arrangements, mechanical or crystal filters etc. may be switched between transmitter and receiver as necessary. The use of a common local oscillator greatly facilitates

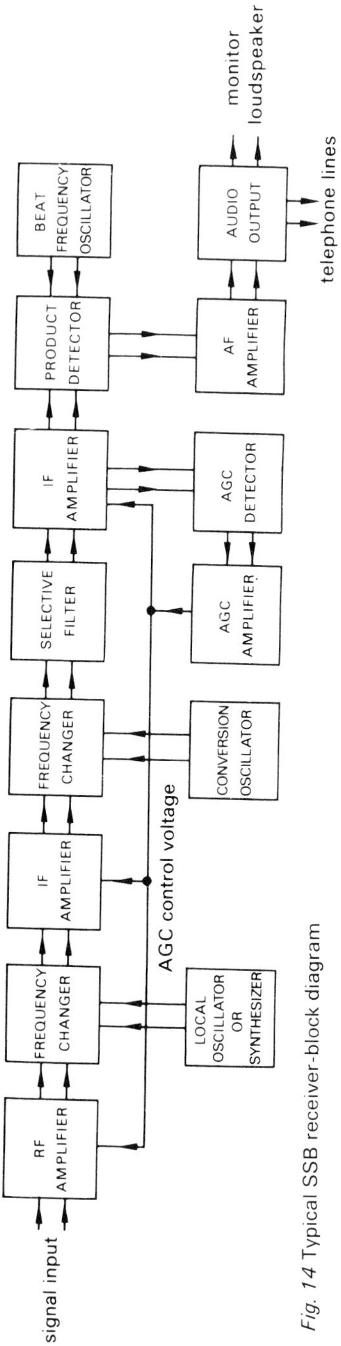

Fig. 14 Typical SSB receiver-block diagram

'netting' and the avoidance of excess duplication of circuitry shows economic advantages.

Airborne equipment

The design differences between ground and airborne equipment are dictated largely by the differing environments. Ground equipment is normally operated on a single frequency, from an incoming mains supply voltage, in a building where the space available makes little constraint on the dimensions or weight of the equipment. On the other hand, airborne equipment is required to operate on all channels within the band from 24 V d.c. or 110 V 400 Hz a.c., remote from the operator's position, often in a confined space with severe weight limitations. Space restrictions on the flight deck limit the avionic presence to control and display equipment with the bulk of the circuitry in standard size boxes mounted in racks in some other part of the aircraft.

The circuitry for both VHF AM and SSB equipment is in general similar to the equivalent ground equipment with the exception that the receiver and transmitter oscillator frequencies are synthesised either as described in the paragraph on SSB receivers or by the selection of a combination of one crystal from each of several banks of crystals, each bank representing one order of magnitude, e.g. crystal from the first bank generates the tens of MHz, a crystal from the second bank, the hundreds of kHz and a crystal from the third bank generates the final tens of kHz. The outputs from the three crystals are combined to develop a submultiple of the required output frequency which is then multiplied by the appropriate factor and amplified to the required output level. Due to severe space and weight restrictions, airborne equipment is extremely compact with a high component density within the individual equipment cases. In consequence care has to be taken with ventilation and allowance must also be made for the fact that the equipment may, particularly in unpressurised aircraft, be operating at an air pressure only a quarter or less of that at ground level.

Multicarrier systems

It is essential that aircraft be able to communicate with the air traffic control authorities whenever flying within controlled airspace or within flight information regions. Unfortunately the range of a VHF radio transmission is limited by the earth's curvature to about four thirds of the optical horizon. This is known as the radio horizon and its distance may easily be calculated from the formula:

$$V = \sqrt{(2h_1)} + \sqrt{(2h_2)}$$

where V = range in miles,
h_1 and h_2 are the heights of the communicating stations in feet.

From this it can be seen that the effective communication range of a radio station to an aircraft below 5000 ft will be less than 100 miles and consequently to cover a country such as the United Kingdom considerable numbers of radio stations would be required. If each station were allocated its own discrete frequency not only would many more channels be required than are available, but the continual need to change frequency on a flight would add considerably to the labours of the flight crew who already have a heavy workload.

In 1947 the UK authorities laid down the principle that as far as possible only one frequency should be used in any airway, sector or Flight Information Region. This solution to the problem within the UK was put into effect with the introduction of the Multicarrier Scheme.

To give the required radio cover it was necessary to operate several adjacent radio stations in parallel on the same nominal frequency from a single control position. If these stations were crystallised to operate on exactly similar frequencies, operational tolerances would cause discrepancies between the stations which could amount possibly to several hundred Hz and aircraft in areas of overlapping cover would in such cases hear a beat note between the two transmissions which could make the speech unreadable. This problem was overcome by arranging for each transmitting station to operate on a slightly different frequency, sufficiently close to the nominal to remain within the passband of the aircraft receiver but with adequate displacement to ensure that any beat note between the transmitters was above the upper limit of the audio frequency response of the aircraft receiver.

With the ex-service T1131 transmitters in use at that time, maintenance of sufficient frequency stability caused considerable problems, for merely adjusting the interstage coupling between the oscillator and its succeeding stage was sufficient to alter the output frequency by several kilohertz. The problem was solved by The Plessey Company who developed a high stability oscillator unit for use with this equipment which remained stable within a few Hertz over a period of many months. Careful circuit design and the use of superior quality crystals mounted in thermostatically controlled ovens ensured that later generations of transmitters were inherently of sufficient stability for the service.

Present practice is that a chain of three multicarrier stations operates on +7.5 kHz, −7.5 kHz and on nominal frequency.

Stations involved in multicarrier operations invariably carry equipment for several channels such as Airways, FIR and Volmet services. These are controlled from the appropriate air traffic control centre which may be a considerable distance from the transmitting station. Control is exercised over landlines and due to the distances involved, repeater amplifiers are frequently necessary. This precludes d.c. switching techniques and requires some form of tone switching. Originally simple on/off tone switching was used, but this has now been replaced by a three state switching tone in which one state will ensure that both main and stand-by equipments are held at a ready state but that spurious noise or tones along the line will not cause either transmitter to radiate. To cause a transmitter to radiate requires that the hold-off state be removed and replaced by one of the other two states, one of which switches the main transmitter and the other, the stand-by equipment.

Common aerial working

For any VHF station operating a multiplicity of channels the number of aerials required is one for each main and stand-by transmitter, main and stand-by receivers. With many channels in use this is obviously impractical so use is made of circuits to permit several receivers or transmitters to use a single aerial.

One of the simplest methods of achieving this aim is the use of a cavity resonator fitted between the receiver or transmitter and a common aerial feeder point. The cavity forms a very sharply tuned circuit which conducts signals at the frequency to which it is tuned but acts as a short circuit to frequencies outside its passband. The filter is connected to the common aerial point by a feeder one quarter of a wavelength long. It is a characteristic of transmission lines that if a short circuit is imposed on the end of a quarter wavelength line then the other end of the line will act as an open circuit. The combined effect of the resonator and quarter wave line will therefore be that, except at the resonance frequency of the resonator, the combination presents a very high impedance at the common aerial point and consequently several equipments may be connected to the common aerial feeder point without interaction. Although by this means it is possible to operate several equipments to the same aerial, the resonator bandwidth is finite and allowance for this must be made in the selection of frequencies.

3.2 Direction finding

The directional properties of the classic inverted 'L' aerial were first described by Marconi who, in 1905, patented a system in which a number of these aerials were arranged radially about a point. By comparing the strength of signal received on each aerial, the direction of the transmitting station could be ascertained.

From that time onwards development was rapid and by 1915 to 1916 the UK had installed a chain of accurate direction finding stations around the coast of the British Isles for monitoring and locating German ground stations, ships and aircraft.

Using Bellini-Tosi or Adcock type aerials, MF D/F stations formed the mainstay of aero navigational services between the two wars (1918–1939) and, although being superseded by more modern aids, remained in service in some areas of the world until the mid 1950s.

The introduction of VHF Radio Telephony in the Second World War led to the development of suitable direction finding equipment for these frequencies. The most common within the UK was the Type 61 homer which, although originally produced for the armed forces, later saw service at almost all British airports. Although now obsolete, a description of the equipment provides a useful introduction to the principles of VHF D/F.

The Type 61 homer

The operation of any D/F equipment depends on the directional characteristics of its aerials and although the older MF systems had simulated a rotating aerial by sampling the outputs from either two crossed loops (Bellini-Tosi) or two pairs of vertical aerials (Adcock) using a device called a goniometer, the introduction of wavelengths of the order of 3 m, for the first time, made the use of rotating aerials a practical proposition.

The most basic form of aerial is the half wave dipole. This, as the name suggests, is approximately half a wavelength in length and split at the centre at which point the feeder is attached. If mounted vertically such an aerial receives equally well from all directions. This may be shown diagrammatically by means of a polar diagram.

A polar diagram shows in convenient form the efficiency of the aerial in any direction either in the horizontal or vertical plane. The two types of diagram are known as the Horizontal Polar Diagram (HPD) and the Vertical Polar Diagram (VPD) respectively.

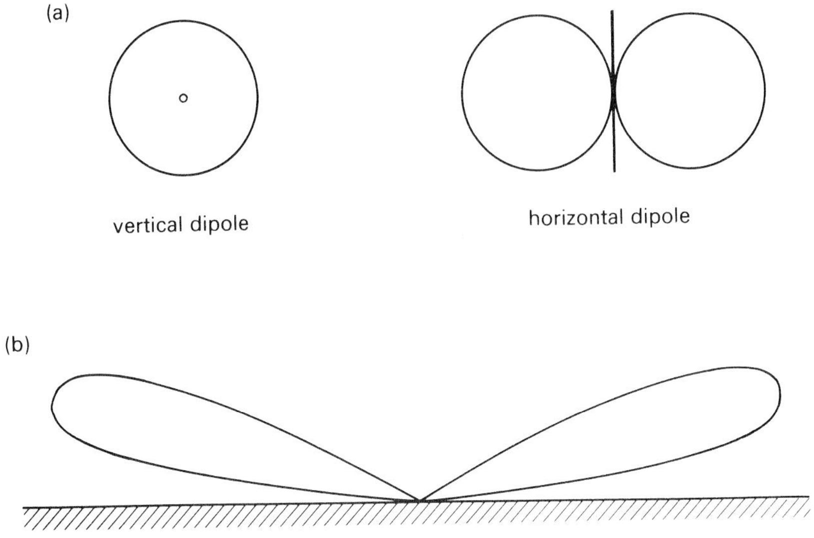

Fig. 15 (a) Horizontal polar diagrams. (b) Vertical polar diagram of a horizontal dipole, one half wavelength high.

The horizontal polar diagram of a vertical dipole is therefore a circle with a dot in centre representing the aerial.

If two such aerials are situated close together, each will effect the efficiency of the other in certain directions and when both are connected to the same equipment, it is then convenient to consider the combined polar diagram. When two vertical dipoles are mounted in the same horizontal plane at a distance of half a wavelength and fed by equal length feeders, such that the upper half of one dipole is in parallel with the lower half of the other, the HPD will take the form of a figure 8 with very sharp nulls. As the aerial is of such a size that it can be turned by rotating the aerial until the received signal disappears, a very accurate bearing of the incoming signal may be taken. However, due to the symmetry of the polar diagram, it is not possible to determine from which side of the aerial the signal is being received. For this reason reflector elements are placed one quarter wavelength behind each dipole.

These considerably improve reception from the side opposite the reflectors and degrade it in the other. A switch is fitted in the centre of each reflector element which if opened renders the reflectors inoperative.

The operation of taking a bearing is then as follows: on hearing a signal, the operator rotates the aerial until the signal disappears and notes that bearing from the calibrated scale. The aerial is then turned a few degrees until the signal is again audible and the 'sense' switch depressed. This opens the

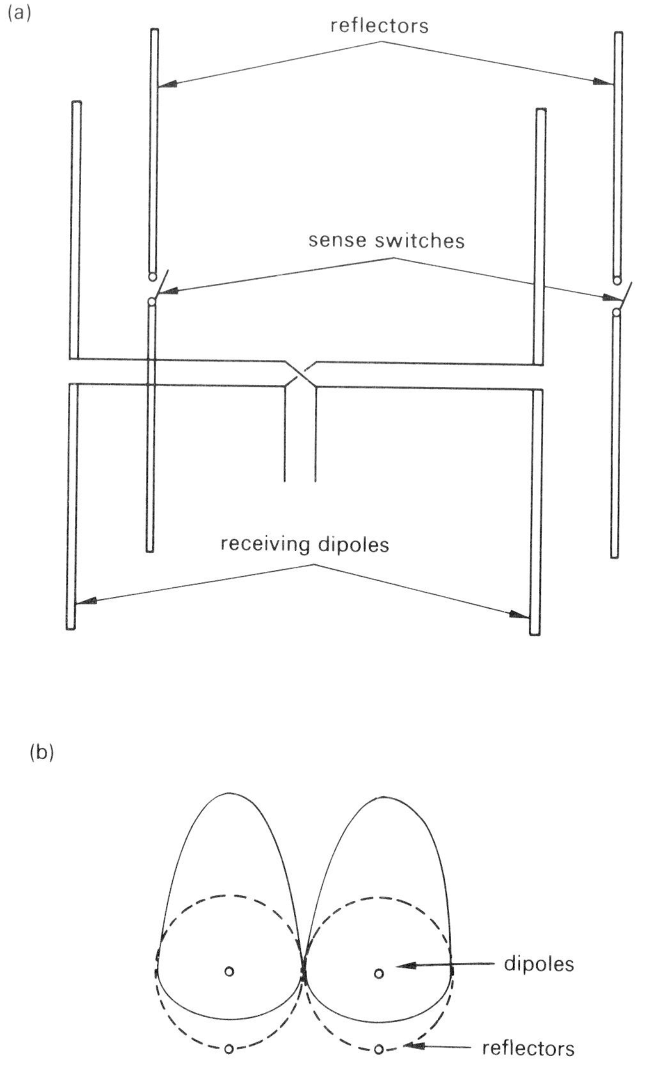

Fig. 16 (a) Type 61 aerial–schematic diagram. (b) Type 61–horizontal polar diagram (broken line indicates sense switch depressed).

switches in the centre of the reflector elements and if the signal decreases in strength the bearing is correct because, in breaking the reflector elements, the gain of the combination is lost. Conversely, if the signal increases on depressing the sense switch, the shielding effect of the reflectors has been lost and the incoming signal is to the rear of the aerial.

The Type 61 homer was capable of an accuracy of $\pm 2°$ and was used for two purposes; for position fixing and as an aerodrome approach aid.

In the position fixing role, during and after the Second World War, numbers of Type 61 equipments were situated across the country, those in any one particular area forming a 'chain'. Bearings were taken on all transmissions heard on a predetermined frequency and passed by telephone to a central point where the incoming information was transferred to a plotting table. By combining bearings received from several stations the position of the transmitting station (i.e. aircraft) could be determined.

The Type 61 gained a very high reputation as an approach aid and for this purpose remained in service on many of the small Scottish airports until the early 1960s, the last one being withdrawn from service in 1962 at Machrihanish Aerodrome, Campbeltown, Argyll. Procedures used by service and civilian pilots varied; whereas in the armed forces it was the responsibility of the ground controller to ascertain course corrections based on information received from the homer operator, civilian ground staff only passed the bearings to the aircraft whose pilot interpreted the information to allow a successful approach to be achieved. In either case the pattern flown was similar.

The Type 61 was normally situated as near as possible to the centre of the airfield to enable it to serve all runways with minimum alignment errors.

Where only one runway was in use, the siting tended towards the mid-point of the runway adjacent to the parallel taxiway. The equipment was calibrated both in true bearing (QTE) and magnetic reciprocal (QDM), the latter being the magnetic compass course that the approaching aircraft would have to steer to reach the airport.

From the time of first hearing the aircraft a bearing is taken on each transmission. These are transmitted to the aircraft pilot who uses this information to facilitate an approach to the airport on a reciprocal bearing to the runway in use. During this approach, height is reduced to a predetermined altitude, usually 3000 ft above airfield level. As the aircraft nears the airport, frequent special transmissions are made to the homer until the aircraft is overhead and the homer reports 'no bearing'. The aircraft then commences descent at 500 ft/min for a period of 3 min using bearings to ensure that a constant course is maintained. A level standard turn is then initiated, and when the extended centre-line of the runway is reached, descent is recommended for 2 min approaching the runway. At this time the aircraft should be at 500 ft with the runway in sight, directly ahead.

Such procedures enabled approaches to be made in meteorological conditions of 800 yd visibility and 500 ft cloud base.

Disadvantages of the Type 61 homer

Although in many ways an excellent aid, the Type 61 had certain disadvantages, not the least of which was the requirement for a specialist D/F operator. Primary radar and instrument landing system replaced it as an approach aid and the Type 61 was relegated to use as a method of identifying radar returns. It was then inevitable that the Type 61 should be replaced by more modern automated systems for which a display could be fitted in the air traffic control desk or radar console.

Automatic direction finders

The Type 61 was replaced by the first generation of automatic direction finders, the most common in the UK being the Marconi AD210. More recently, these, in turn, have been replaced by equipment using the Doppler principle. These show considerable advantages over the earlier systems, for the principle of operation is such that the determination of bearing is independent of the signal strength of the received transmission. This, combined with the wide aperture of the receiver aerial array, permits a far greater accuracy than was formerly possible.

Doppler direction finding systems

The Doppler effect is well known and is perhaps most commonly noticed as a drop in pitch of the sound of a passing vehicle. To the stationary observer the sound is apparently raised in pitch as the vehicle approaches and lowered as it recedes, the faster the vehicle, the more pronounced is the effect.

This effect is common to all wave motions including radio waves. If the frequency of a transmission from an approaching aircraft is accurately measured from a stationary point on the ground, it will be found to be higher than that radiated from the aircraft, but when receding it will be lower. Only at the point of nearest approach (i.e. when the aircraft is neither approaching nor receding) will the frequency transmitted and the frequency received be the same. In practice this frequency shift is insufficient to cause any reception problems but it is still sufficient to be the basis of the principle used in many modern direction finders.

Consider an aircraft transmission being received on a simple vertical aerial. The received frequency will be that radiated from the aircraft plus or minus a few hertz due to Doppler Effect consequent to the motion of the aircraft. However, during the period over which a bearing is taken, it is unlikely that the velocity of the aircraft will vary significantly and the frequency received by the aerial may be considered a constant.

If the aerial is moved rapidly, Doppler effect will again be evident, movement towards the aircraft causing an increase in the received frequency and movement away, a drop.

Now consider changing the movement of the aerial to a path described by the circumference of a circle. The frequency of the incoming signal will then vary sinusoidally as the aerial rotates around the circle and if the rotation is at constant speed, the incoming signal will be frequency modulated at the frequency of aerial rotation, the phase being dependent on the relative bearing of the aircraft and receiving station.

To measure the phase of the modulation a reference signal is necessary.

This may be conveniently produced by attaching a generator, whose output is one cycle per revolution, to the shaft around which the aerial rotates. A phase comparison between this and the frequency modulation on the received signal will give the bearing of the incoming signal. To achieve adequate system accuracy, it has been found that the moving aerial must describe a circle in the order of 8 m diameter at a rate of between eight and twenty times per second. In mechanical terms this degree of movement is not practical within the operational environment. The moving aerial has therefore been replaced by a circle of static aerials, which, if each is sampled in turn, will give an overall result equivalent to a single rotating aerial. This commutation is achieved by electronic switching, with the reference signal being derived from the switching waveforms.

After phase comparison the positional information may be displayed in a variety of ways. These include display on a miniature CRT, a radar PPI, a mechanical meter or on an electronic digital display.

Manufacturers claim bearing accuracies of better than ± 1° for this type of equipment, this being due to the combination of the accuracy of the phase comparison techniques, and the relative immunity to siting errors inherent in wide aperture aerial systems.

Fixer systems

The directional information provided by a single direction finding installation is sufficient to enable a successful approach to an airfield to be achieved, but at no time can the position of the aircraft be derived from the D/F derived data alone. Such an approach is made using compass and time in addition to the D/F bearings which are co-ordinated to enable the required approach pattern to be realised. The determination of the position of an aircraft or vehicle by direction finding equipment alone requires the services of several D/F stations, the bearings from which must be relayed to some central point where they may be plotted on a suitable map, the intersection of the bearing lines from the individual stations indicating the position of the transmitting aircraft or vehicle. A series of stations operating together in this manner is known as a 'fixer' chain and such chains have been in service from the early part of the First World War. The efficiency of this system, even in those days, was testified by the German Zeppelin captains who considered it to be more accurate than any other form of navigation for operations over the UK.

Between the two world wars many fixer chains were operational on the MF and HF aeromobile bands, some remaining in service until well after the Second World War. During the Second World War the RAF established an extremely efficient VHF chain using Type 61 installations which remained in service for many years. With the introduction of more sophisticated aids such as VOR, the need for a fixer service for en-route navigation diminished and the service was retained on the emergency frequency only (121.5 MHz). For this role, more modern automatic equipment was installed and the service extended to cover the UHF emergency channel (243 MHz).

The use of automatic direction finding equipment led to alternative and more speedy methods of plotting. Whereas with the manual direction finders the measured bearing was relayed by voice along telephone lines and manually plotted, with the more modern equipments the bearing information could be transmitted directly to the control centre and displayed by means of a projection cathode ray tube or laser projector.

A projection cathode ray tube is similar to a normal CRT except that by virtue of operating at extremely high voltages and the selection of special screen phosphors, it displays an exceptionally bright picture. This picture is focussed by an optical system to provide a greatly enlarged image on a distant screen.

Within the air traffic control centre, the plotting table consists of a vertical, semi-transparent map of the airspace covered by the fixer chain. Behind the map are a number of optical units, each consisting of a projection CRT and optical system focussed on the rear of the map. Each optical unit is connected to the output of a single automatic D/F station and the position of the projected spot, when no signal is present, is arranged to correspond with the geographical position of the associated station on the map. When a transmission is received, the spot is deflected into a line which indicates the bearing of the transmitting station relative to the direction finder. As any signal will be received by more than one station, at some point the projected lines will intersect, this being the position of the transmitting aircraft.

By such means, the Distress and Diversion Controller viewing the face of the map, will immediately be aware of the position of the aircraft requiring assistance, thus greatly facilitating his task of rendering aid.

3.3 Siting of direction finders

The accuracy of any radio navigational aid is dependent to a greater or lesser degree on siting. Although almost all equipment developed over the past thirty years have been capable of extremely accurate results on very good sites, their performance under the frequently far less than ideal circumstances dictated by operational need is of far greater interest to the telecommunications engineer. This has led in recent years to the development of 'wide aperture' systems which tend to be less affected by siting restrictions than earlier equipments.

Ideally a direction finder should be situated on a high flat plateau with no obstructions in any direction, but in practice it is almost certainly on or adjacent to an aerodrome with consequent obstructions such as control towers, water towers, hangars and freight sheds. It is dependent on the skill of the siting engineer to ensure that the effect of these obstacles is minimised.

The cause of all errors attributable to siting is the reception of indirect radiation. This can reach the aerial in three ways, by reflection from large objects in the field of the transmitter, by re-radiation from conductors near the receiving aerial and by radiation diffracted around large obstacles between the transmitter and the receiver. Of these the first is by far the most common.

This effect is most evident when a large object such as a hangar lies a little offset from the direct path between transmitter and receiver. The signal from the aircraft is omni-directional, therefore not only will it impinge on the receiving aerial but also on all objects within range of the transmitter. When 'hit' by the transmission the object will reflect a proportion of the signal which will be received by the D/F aerial but from a different bearing to the direct path transmission. The direction finding equipment therefore finds itself presented with a signal arriving simultaneously from two different directions. Under these circumstances the automatic direction finder frequently gives an indication somewhere between the correct bearing and that of the reflecting object, depending on several factors which include the relative strength and phase of the two signals. The efficiency of any object as a reflector is extremely difficult to forecast as it varies with operational frequency, constructional material, as well as the shape and angle of incident radiation. The D/F equipment should therefore be sited as far as possible from such reflectors so that their effect be minimised and that the ratio of received direct to indirect signal be maintained as high as possible.

Topographical features may also give similar effects. A large hill or mountain, for example, may well cause large bearing errors in a particular sector even if it is several miles away. If this is so the situation can only be alleviated by raising the height of the VDF equipment, e.g. by siting on a convenient local hill.

Near field conductors

Conductors near to the direction finder can also have a considerable effect on bearing accuracy. Any conductor in the field of an electromagnetic wave will re-radiate a proportion of the incident energy. As VHF aeromobile signals are vertically polarised, vertical conductors will have the most effect. Fortunately the strength of the re-radiated signal diminishes rapidly with distance and typically a 10 m steel tower (e.g. a wind pump) at 200 m and a fence with 1.5 m steel posts (but not wire mesh) at 100 m would be acceptable. Horizontal conductors such as telephone wires or power lines may also cause errors, particularly when the signal being received is from high flying aircraft. To minimise this effect the D/F should be sited such that telephone wires (on wooden poles) do not approach nearer than 200 m and power lines (on steel pylons) nearer than 1 km. For Super-Grid pylons (which are much larger) the distance must be much greater.

Refractions from distant mountains

This is rare and normally will only affect low flying aircraft in mountainous regions. When the aircraft is shielded from the direction finder by a mountain one signal will be received by diffraction over the mountain whilst other signals are received by diffraction round the sides of the mountain. Little can be done about this other than to situate the D/F as high as possible to minimise the shielding effect and to inform the pilot of the possible magnitude of the error, e.g. Class A (±2°), Class B (±5°), or Class C (±10°) bearing.

General siting rules

Although the ideal site as described at the beginning of this chapter is usually unattainable, every effort should be made to ensure that distant hills do not subtend an angle of above 2° or 3° above the horizontal. It is sometimes possible to waive this requirement in the case of an isolated peak if an impaired performance can be tolerated in one or two directions.

Closer to the aerial, i.e. within a 100 m radius, the ground should be clear of bushes and trees etc. with irregularities not exceeding ±15 cm. Outside this area isolated small obstructions are permissible, increasing in size with distance. The general principles for the siting of direction finders is equally applicable to all other VHF navigational aids.

Section 4

Short range navigation and approach aids

4.1 Non-directional beacons and automatic direction finders

The simple medium frequency non-directional radio beacon (NDB) operated in conjunction with the airborne direction finder is possibly the most common and, from the ground engineer's point of view, the simplest navigational aid in use today.

The ground equipment

The ground equipment is a conventional MF transmitter operating on a frequency in the 200 kHz to 500 kHz band, which radiates an uninterrupted carrier modulated at regular intervals by a tone keying the callsign of the beacon in international morse code.

A non-directional beacon (NDB) may be used in the vicinity of an airport as aerodrome or ILS locator, or in association with VOR or alone on airways. Recently an NDB on an airfield has been teamed with a DME to provide a simple approach aid.

The power output of the equipment is dependent on the range required for the service being operated. An airways NDB could be feeding several kilowatts to a highly efficient aerial to achieve a range in excess of 200 miles, whilst for an aerodrome locator beacon it may well be found that ten watts to a 10 m whip aerial are sufficient for a required 10 mile service area.

The frequency band in which NDBs operate is extremely congested and inevitably close tolerances in terms of both frequency and output power are necessary.

The frequency stability of an equipment is a function of its design and of the quality of the crystal in the frequency determining circuit. In present-day equipment this aspect presents few problems.

The field strength required of a beacon at the limit of its coverage depends largely on the noise level in the area concerned. Whereas, in areas which suffer only relatively low noise levels such as Europe, 70 microvolts per metre are adequate, in hotter climes, between latitude 30° North and 30° South, 120 microvolts per metre or even higher may be necessary. When the power level necessary to achieve the required coverage has been ascertained, ICAO regulations require that power output of the beacon should not exceed this level by more than 2 dB.

The identification of most beacons is effected by modulating the carrier with a tone of either 400 Hz or 1020 Hz, keyed with the station identification letters in international morse code. To ensure maximum intelligibility, the modulation depth is kept as high as possible although the depth of modulation possible is not always set by the transmitter itself. Short range beacons frequently use aerials which in terms of the frequency in use are extremely short and are therefore highly reactive and tend to tune extremely sharply. In extreme cases this can mean that although the aerial is accurately tuned to the carrier frequency, the sideband frequencies are sufficiently off resonance to be severely attenuated. Tests have shown that in extreme cases modulation depth can be limited to 50% by this effect.

To ensure uninterrupted service the usual practice is for two transmitters to be installed at each beacon site. Also included is a monitor system which analyses the RF signal for signal strength and modulation depth. Should either fall below predetermined limits (normally 3 dB) a changeover sequence is initiated, closing down the operational transmitter and restoring service with the stand-by equipment.

Aerial and earthing systems

At the frequencies used for the beacon service the wavelengths vary between 1500 m (200 kHz) and 600 m (500 kHz) thus it is generally impractical to erect an aerial approximating to a resonant length. Recourse is therefore made to erecting an aerial of convenient length and inserting additional inductance and capacity to achieve resonance. This process is known as 'loading'. Typical aerials in use vary from 'T' aerials 25 m high and 50 m long for long range beacons to 10 m towers insulated from ground and 'whip' aerials varying from 10 m to 20 m in length. Recent designs of the latter type

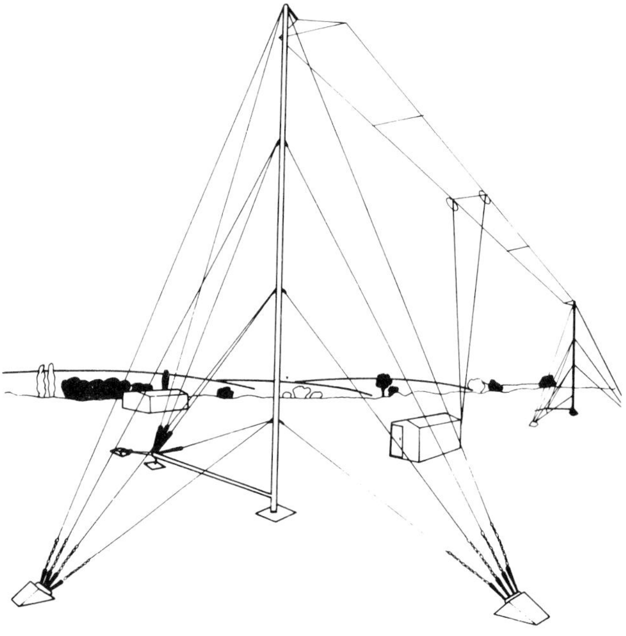

Fig. 17 NDB 'T' aerial system. (By courtesy of Decca Navigator Company Ltd)

of aerial include an insulated section at approximately the mid-point across which an additional loading coil is connected, giving rise to a noticeable rise in efficiency.

Paradoxically, one of the most important parameters affecting the efficiency of a short aerial system is that of the earth system. Short aerials, such as used in non-directional aerials exhibit extremely low radiation resistance. The earth resistance is effectively in series with this. Consequently the earth resistance must be small compared with the radiation resistance of the aerial to ensure efficient radiation.

The earth resistance depends upon the extent of the earth system, the nature and the moisture content of the soil, the latter being the factor most susceptible to change. When the radiation resistance of the aerial and the earth resistance are comparable, such a variation, due for instance to a change in weather conditions, can cause a considerable difference in the signal strength radiated with consequent difficulties in maintaining the required service area. Two reasons compound to give this effect. In so far as the transmitter is concerned, the aerial and earth are effectively a single circuit and the output power is divided in proportion to their relative resistance. A variation in either radiation or earth resistance will therefore cause a change in the power distribution with consequent effect on the radiated signal. Furthermore, the variation in circuit constants may well cause a mismatch to occur between transmitter and aerial thus still further eroding the output of the transmitter.

To minimise these effects it is usual to install the most effective earth system possible. Typically this consists of a number of radial wires under the aerial extending outwards to a distance equivalent to the height of the aerial and terminated in a spike driven several feet into the ground. These wires are frequently made of lead to minimise corrosion problems and take the form of 1 in wide by $\frac{1}{4}$ in thick strips. Alternatively, metal matting covering a similar area can be used, but simple earth spikes alone are rarely adequate except where earth conductivity is extremely high.

The airborne equipment

Although the principles of the design of ground MF beacons have remained virtually unchanged since the inception of the system, the associated airborne system has been the subject of continuing development.

In the earliest days a fixed loop aerial was installed in the aircraft positioned such that the nulls pointed fore and aft. This enabled aircraft to fly radial courses to and from beacons but suffered from complexities caused by cross wind. This causes the aircraft to 'crab' so that it no longer points its nose directly towards its destination. This system was replaced by the rotatable loop which came into widespread use in the 1930s but was later abandoned because of the additional workload its operation imposed on the flight crew compared with its successor, the radio compass.

The radio compass makes use of a rotating loop aerial driven by a motor which, on reception of a signal, automatically rotates the loop until a null is found. A selsyn system displays the bearing on a mechanical instrument in the cockpit. The earlier systems used a loop aerial with about a 9 inch diameter installed either in a tear-drop shaped fairing about a foot away from the aircraft skin or in a suitably sized dome projecting into the slipstream. Since the mid-1950s, however, the development of aerials wound on ferrite slabs has enabled the production of efficient loop aerials only an inch or two deep which need only project a few inches above the aircraft skin.

The most recent equipment use a fixed cross loop system working in conjunction with a motor driven goniometer. A major advantage of this system over those using rotating loops is that all moving parts are contained within the receiver box. Antenna projection from the aircraft skin can be as low as 1 inch with a horizontal dimension of about 1 ft.

The accuracy of any of these systems is in the order of 2° exclusive of errors induced by the aircraft structure. These errors are frequently of considerable magnitude except in the fore-and-after directions where they tend to be minimised due to the symmetry of the airframe. Some errors may be corrected for a given aircraft type but this correction will only be for one frequency and one aircraft attitude. In consequence airborne direction finders cannot normally be relied upon to better ± 5° when receiving ground wave signals. When sky waves are present the possible error increases to ± 30°.

The cockpit display indicates the bearing of the received signal compared with the centre line of the aircraft. Whilst the earlier indicators consisted of a single needle traversing a circular scale calibrated 0–360 degrees, more recent equipments use an instrument known as a radio magnetic indicator. This displays separate indications from two ADF sets, two VORs or one of each. The outer scale also rotates independently, being synchronised to the gyro compass.

Operation of the airborne equipment

By nature of the physical restrictions inevitably placed on MF aerials, their radiation tends to be predominantly vertically polarised. In such a transmission the direction of action of the electric component of the wave is vertical and that of the magnetic component, horizontal. Both components act at right angles to the direction of travel of the field.

The loop

If an inductive aerial consisting of one or more loops of wire is introduced into such a field it will couple with the magnetic component of that field. By turning such an aerial about its vertical axis, the amount of coupling may be varied. At the point of minimum coupling it can be used to indicate the line of action of the magnetic component and therefore the line of travel of the electromagnetic field. If the loop is rotated through 360° it will be found that the polar diagram corresponds to a figure of eight.

At any instant, the voltage induced in the loop is proportional to the rate of change of the density of the magnetic flux coupling with the loop. As this magnitude varies sinusoidally with time, so therefore, does the rate of change of flux density vary similarly, but its phase leads that of the magnetic variation by 90°. Consequently, the phases of the voltages at the loop terminals will either lead or lag that of the electromagnetic field according to whether the position of the loop is in its first or second 180° of rotation.

The sense aerial

As the electric component of the electromagnetic field acts in a vertical direction, it is independent of the direction of travel of the field. A capacitive aerial may therefore be fitted to the aircraft which will couple only with the electric component of the field. This is known as the 'sense' aerial and its output voltage is developed between the base of the aerial and the aircraft structure.

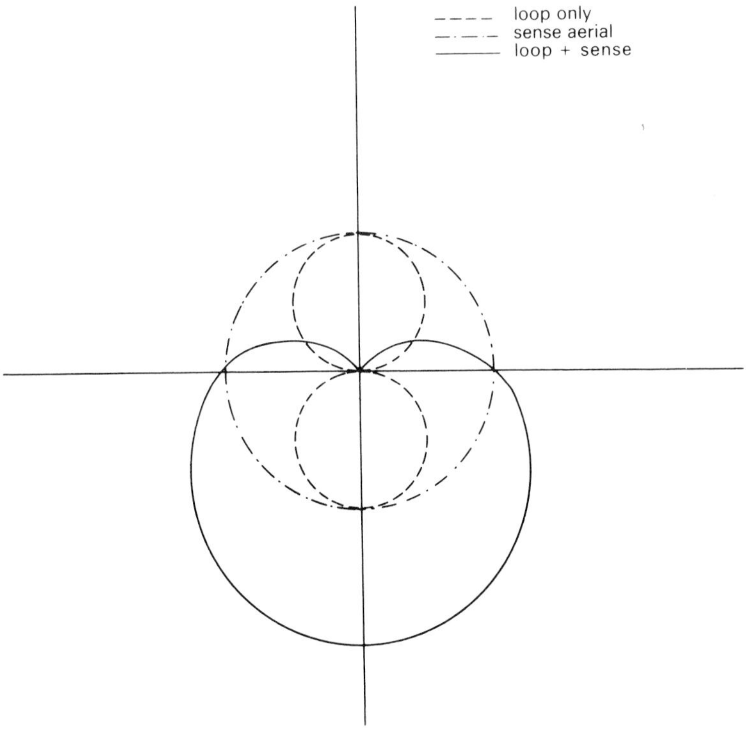

Fig. 18 Loop aerial polar diagrams

The ADF receiver

In a typical ADF receiver the phase of the loop voltage is advanced by 90° so that the signal is 180° either in or out of phase with the electromagnetic field which is acting on both loop and sense aerials. This signal is coupled to a phase splitter which feeds a balanced modulator consisting of two electronic switches. These are operated alternately by a low frequency switching waveform to release 'blocks' of loop signal which are alternately 180° in and 180° out of phase with the magnetic field.

These 'blocks' of signal in conjunction with the output of the sense aerial are then applied to an amplifier. The two signals combine, adding or subtracting depending on relative phase. The output of this amplifier is therefore an RF signal modulated at the switching frequency whose modulation depth is dependent upon the relative magnitudes of the sense and loop signals at the common input and whose phase, compared with the switching waveform, indicates the attitude of the loop aerial.

This output is then phase shifted a further 90° and used to control the magnitude and phase of the power supplied to one coil of a two phase motor which serves to rotate the loop aerial.

In modern high speed aircraft, aerodynamic factors frequently inhibit the use of rotatable ADF loops. In such cases a 'fixed loop aerial' is fitted. This device consists of two small aerial coils wound on ferrite cores mounted at right angles. These are connected to a goniometer fitted within the receiver. The goniometer search coil is driven by bi-phase motor controlled in a similar way to the rotatable loop described previously.

4.2 VHF omni-range beacons (VOR)

VHF Omni-Range (VOR) is the internationally recognised short range navigational aid. It is used, normally in conjunction with DME, to delineate airways and a low power version is frequently used on small airports as an approach aid. Developed in the USA from the rotating beacon of the 1920s, it was recognised as an international standard in 1949 and has subsequently replaced the MF radio range.

The principle of operation is that two independent 30 Hz modulations are impressed on a VHF ground station radio transmission in the 112.0 MHz to 117.9 MHz band. These two modulations are known as the reference and variable phases and their difference in phase, measured in degrees, as received at any remote station, corresponds to the bearing of that station with respect to magnetic north.

The variable phase is a 30 Hz amplitude modulation whilst the reference phase consists of a 30 Hz frequency modulation impressed on a 9960 Hz amplitude modulated sub-carrier. In older equipments the variable phase is generated as a space modulation but on the more modern 'Doppler' equipments a pseudo-Doppler effect generates the FM on the sub-carrier of the reference phase whilst the variable phase is a conventional 30 Hz amplitude modulation.

Advantages of the VOR over the older MF ranges include freedom from night effect and sky wave errors whilst guidance is increased from four delineated courses to navigational information at all bearings.

The major errors suffered by the system are due to siting (reflecting objects near the transmitter) and errors in measurement of the 30 Hz phase differences in the airborne equipment.

Horizontal aerial polarisation was selected after comparative tests showed less susceptibility to site errors despite the fact that this complicated the aerial systems.

Conventional VOR systems

As is so often the case, the principles of the system can most easily be understood from a description of the earliest types of equipment.

The required output signal from a VOR consists of reference and the variable phase signals on the same carrier wave, identification signals in morse code at regular intervals and speech modulation, although the latter is rarely used within the UK.

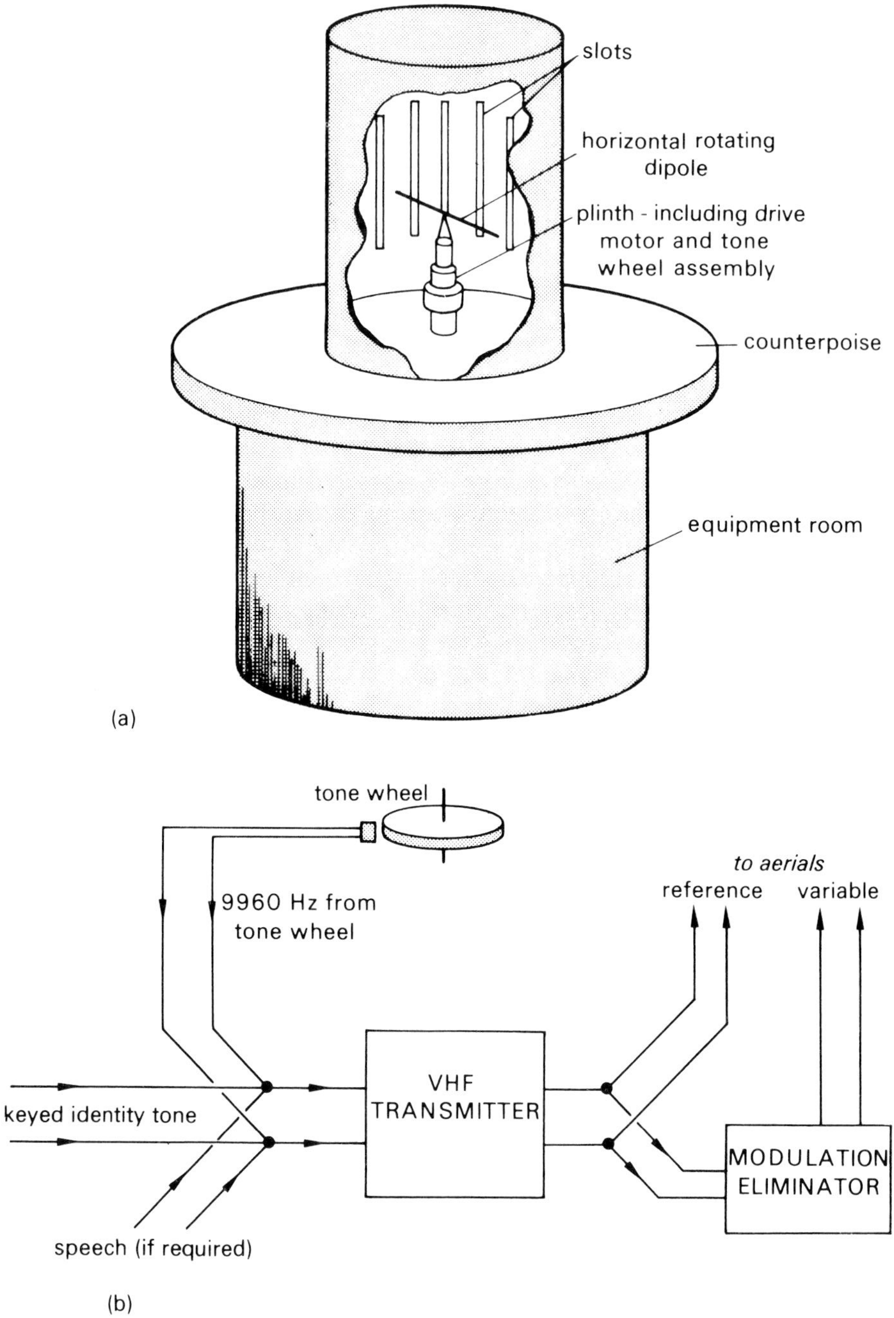

Fig. 19 Conventional VOR (a) Building layout. (b) Schematic.

The transmitter

The transmitter is an AM transmitter of standard design except that the modulator must be capable of operation at frequencies of up to 10 kHz. For en-route aids the output power is 200 watts but for airfield service, in which case it is called a Terminal VOR or TVOR, only 50 watts are necessary.

The transmitter modulator receives input from three separate sources: the reference phase generator, the identification signal generator and from a speech circuit.

The most basic form of reference phase generator consists of a metal wheel mounted on a shaft and which has 332 teeth cut around its circumference. Mounted closely adjacent to these teeth is an electromagnetic pick-up which gives an electrical output each time a tooth passes by its head. If the wheel is rotated at 1800 rpm the output will therefore be at 9960 Hz, i.e. the reference sub-carrier frequency. The teeth, however, are arranged in a somewhat staggered manner and this irregularity imparts a cyclic variation of between 9480 Hz and 10 440 Hz on the output frequency, the actual frequency being dependent on the instantaneous position of the toothed wheel. This circuit is commonly called a tone wheel generator.

It is obviously of prime importance that a pilot receiving automatic navigational information should be able to identify the source of that information in order to ascertain his position.

Identification of all navigational aids is therefore achieved by incorporating an audio tone, keyed to form the identification letters of the station in international morse code.

The tone is generated by a simple tone oscillator operating at the desired modulation frequency which is keyed in older equipments by a mechanical keying device and in recent equipments by a solid state unit using logic circuitry.

Allowance is also made in the modulator design for the input of speech. This may typically take the form of pre-recorded meteorological, or air traffic control (e.g. aerodrome serviceabilities etc.) information.

The output from the transmitter is split two ways in the approximate proportions 75% to the aerial system (reference phase) 25% to a modulation eliminator circuit.

The purpose of the modulation eliminator is to strip the modulation from a portion of the transmitter output signal to provide the unmodulated RF power for the variable phase signal. The advantage of this system for providing the variable phase signal power is that should the transmitter output power vary, the amplitude and phase relationship between the reference and variable phase signals will remain constant and thus the bearing information will remain unaffected.

The aerial system

The heart of the VOR beacon lies in the aerial system. This can take several forms, one of the most common, and successful, combining the properties of vertical slot aerials and a rotating horizontal dipole.

The equipment is normally situated in a single storey building surmounted by the aerial system. This consists of a solid metal or mesh counterpoise system 20 ft to 40 ft in diameter at the centre of which is situated a metal 'dustbin' shaped structure about 10 ft high and 6 ft in diameter.

A number of vertical slot aerials are cut around the side of the 'dustbin'. These are fed in phase from the reference phase output signal and give a substantially omni-directional radiation pattern.

It has been found that if a slot of suitable dimensions is cut in a sheet of metal and an RF signal applied across that slot, it will act in a similar way to a solid radiator except that a vertical slot radiates horizontally polarised signals and a horizontal slot radiates vertically polarised signals.

The variable phase signal from the transmitter is fed to the rotating dipole. The horizontal polar diagram (HPD) of a horizontal dipole aerial is a figure of 8 and although this aerial is located within the 'dustbin' the HPD is unaffected as any signal radiated from the dipole will be received and re-radiated by the slots. When combined in correct amplitude and phase with the signal being radiated from the omni-directional aerial the resulting HPD becomes a limaçon, a somewhat 'heart-shaped' figure. This polar diagram has the characteristic that if rotated by turning the aerial the signal strength at any remote receiving station will vary sinusoidally. Thus if the dipole is rotated 30 times per second, a receiving station will receive a signal varying sinusoidally in strength 30 times per second, i.e. modulated by 30 Hz.

The rotating dipole and the tone wheel are mounted on a common shaft and consequently the phase relationship of the resulting signals remains constant even if slight alterations in motor speed cause frequency variations. To minimise these variations, the aerial/tone wheel assembly is driven by a synchronous motor fed from a high frequency stability source as the public electricity supply is not considered to be sufficiently stable. This source may typically use a tuning fork or high stability oscillator as its frequency determining element.

The phase relationship between reference and variable phases is arranged to be such that they are coincident at magnetic north. Any aircraft station on this bearing will receive both signals in phase. At any other bearing, due to the time taken for the aerial – and consequently the limaçon – to rotate, the phase of the 30 Hz component of the signal received from the dipole will lag behind the reference phase. As the modulation frequency of the reference phase is equal to the angular rotational speed of the dipole radiating the variable phase, the difference in phase between the two signals, measured in degrees, is equal to the bearing of the receiving station from the VOR.

Monitoring

Monitoring of the radiated VOR signal is achieved by receiving the signal on a small aerial situated on the counterpoise edge or on a horizontal dipole located on a pole some 50 ft from the aerial system. A simple diode detector is incorporated in the monitor aerial system and the demodulated signal is fed back to the executive monitor within the transmitter building. Here, after suitable amplification the signal strength, the levels of variable phase and subcarrier, and phase relationship between reference and variable phase are measured, any variation outside preset tolerances causing the monitor to initiate transmitter switch-off and stand-by equipment to take over service.

Alternative systems

The method of generating the VOR signal as previously described is not unique, consequently most manufacturers depart from the standard circuit in major or minor ways.

The STC VOR of the early 1960s utilised a modified tone wheel system. In this equipment, the metal wheel was replaced by a transparent disc upon which was inscribed radial lines corresponding in position with the teeth on a tone wheel. A lamp was arranged to illuminate a photo-electric cell through the disc, the output of which was fed to the modulator of the transmitter. The rotation of the disc causes the inscribed radial lines to pass between lamp and photocell, thus modulating the light reaching the cell and generating the necessary subcarrier frequency.

A VOR developed by the Wilcox Corporation dispensed with the rotating dipole. In this equipment the variable phase signal from the modulation eliminator is fed to a capacitive goniometer which modulates the signal at 30 Hz and gives two separate outputs which are equal in amplitude but differing in phase by 90°. Each output is split by an RF bridge and energises two slot aerials. The four slots are situated around a vertical tube some 18 inches in diameter, each bridge feeding diametrically opposite aerials.

The output from the four slots interact, producing a rotating figure of eight polar diagram which requires an additional omni-directional signal to develop the limaçon pattern. The reference signal is applied to both aerial bridges, thus energising all slots simultaneously and generating a substantially circular polar diagram.

The Wilcox Corporation also developed an alternative method for generating the reference signal. The outputs from two oscillators, one of fixed frequency and the other tuneable by means of an additional section on the capacitive goniometer, are heterodyned to produce the modulation frequency. The design of the variable capacitive section is such that as the goniometer rotates, the frequency of the associated oscillator varies in a cyclic manner, thus producing the required reference modulation locked to the rotation of the aerial pattern.

Fully solid state VORs

The VOR equipment so far described has incorporated mechanical elements (rotating dipole, goniometer, etc.) as an essential part of the generation of the radiated pattern. However, such elements inevitably suffer from wear and tear, with consequent signal degradation, and in the course of time replacement will be necessary. It is not surprising therefore that considerable efforts have been made to develop fully solid state equipment.

The basic principle of generating the required radiation pattern is that the output from the RF exciter is separated into three paths. The first of these is amplitude modulated by the 9960 Hz sub-carrier which itself is frequency modulated by the 30 Hz reference modulation to a deviation of ± 480 Hz. This is then fed to an omnidirectional, horizontally polarised aerial, typically an Alford loop.

The second and third paths are each fed to a balanced modulator circuit in which they are amplitude modulated by 30 Hz and the carrier suppressed, the only difference between the two being that the phase of the modulating signal varies by 90°. The output from each of these then feeds a pair of vertical slot aerials, the two pairs of slots being positioned such that their maximum radiation is at right angles.

The effect of this is to generate a rotating sideband pattern which, when combined with the carrier radiated from the omnidirectional aerial, produces the variable phase signal.

The accuracy of such a system is obviously totally dependent on the maintenance of the necessary phase relationships. Standard Elektrik Lorenz, in their VOR 4000 equipment, have achieved this by storing all required waveforms in digital form in random access memories. These are then read and transferred to a digital to analogue converter which in turn drives the respective modulation stages.

The above description is obviously very simplistic and, in practice, comprehensive microprocessor-controlled systems are included to monitor all transmission parameters. These circuits frequently incorporate a feedback function in order that any parameter varying from nominal may be automatically corrected.

Should the variation exceed the ability of the feedback circuits to correct (eg. due to circuit malfunction), alarms will be energised and stand-by equipment will take over the operational service.

Site errors and aerial alignment

With conventional VOR equipments the major bearing errors are attributable to siting conditions. The location of a VOR beacon is frequently determined by other than electronic considerations and consequently the philosophy of 'what can't be cured, must be endured' must be accepted. On far from ideal sites, which may well have been necessitated by factors such as the position of airways etc., bearing errors due to ground reflections from uneven terrain, buildings and so on, may be as high as 15° in some sectors. In such cases the aerial is aligned to present the minimum error in the most important sectors. For a beacon delineating an airway this obviously corresponds to the alignment of that airway. At an airway intersection the alignment may be either along the more important airway or for minimum mean error. A Terminal VOR on an airport will be aligned for minimum error along the main instrument runway.

When two or more VORs are used to delineate an airway, particular care is taken in the alignment of their radiation patterns. This is to ensure that on switching from one beacon to the next, the pilot of an aircraft finds the second beacon indicating a course exactly 180° removed from the first.

Any major bearing errors on a VOR installation are notified in NOTAMS (NOtices To AirMen) and also within the UK in the *Air Pilot*, but even so, corrective steps must be taken. The most obvious action is to investigate the possibility of removing the beacon to an alternative site with possible realignment of the airways involved; however, the development of wide aperture VOR equipments operating on the Doppler principle has shown that such equipments are capable of giving considerable reductions in site error.

Doppler VOR

The Doppler principle has been discussed previously in the chapter on direction finders and where used in Doppler VOR is similar except that it is used for transmission instead of reception.

In comparing a Doppler VOR with a conventional system it will be found that the role of the omni-directional aerial and the array are reversed. Phase relationships remain the same, however, so that standard airborne receivers operate satisfactorily with either system.

The aerial system consists of a central omni-directional aerial surrounded in most cases by a circle of 52 aerials 44 ft in diameter. Aerials are usually Alford loops and are mounted above a large metallic mesh counterpoise.

The central aerial radiates an omni-directional signal, amplitude modulated at 30 Hz, this being the equivalent of the variable phase transmission of a conventional equipment.

In the most basic form of the equipment, the aerials in the circle are fed from a separate continuous wave transmission 9960 Hz displaced from that energising the central aerial. Each aerial is energised in turn from a capacitative commutator or solid state switching device to simulate the rotation of a single aerial. As in the Doppler direction finder, when the antenna appears to move towards the aircraft the frequency increases and as it recedes the frequency decreases. It should be noted at this point that in order to maintain the same phase relationships which exist in conventional equipment the apparent rotation must be in the opposite direction, i.e. anti-clockwise, instead of clockwise.

The dimensions of the outer ring of aerials are critical. A diameter of 44 ft combined with a rotational speed of 30 revolutions per second gives a radial velocity in the order of 4150 ft per second. This will cause a maximum Doppler shift of 480 Hz–that required by the specification for the VOR

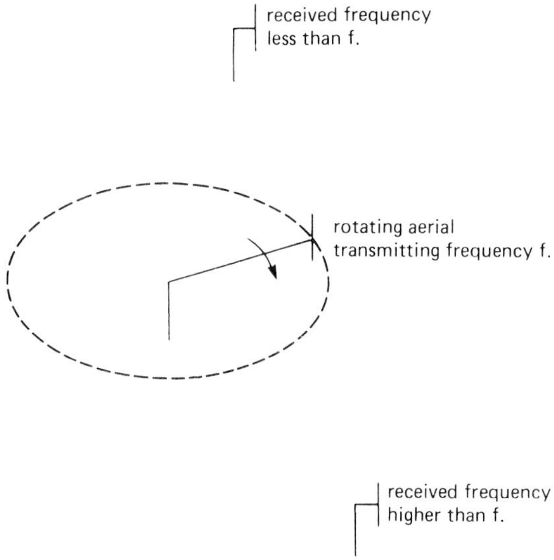

Fig. 20 Effect of rotating the transmitting aerial in a circle. Frequency f will only be received at either receiving site when the transmitting aerial is either at point of nearest approach to, or farthest from, the receiving site

system. The 9960 Hz frequency difference is therefore varied by ±480 Hz at 30 Hz rate with the phase dependent on the bearing of the aircraft.

To the receiver, the signals emanating from a Doppler VOR carry the same information as those from conventional equipments, however, some receivers react unfavourably to operation with a single sideband 'reference' signal, and in consequence modern Doppler VORs are designed to radiate double sideband transmissions. This is achieved by providing separate transmitters for each sideband (i.e. f_c + 9960 and f_c − 9960) and applying their transmissions to diagonally opposite aerials. The frequencies of the variable phase transmission and those of the sidebands are compared to ensure that their correct frequency relationship is accurately maintained.

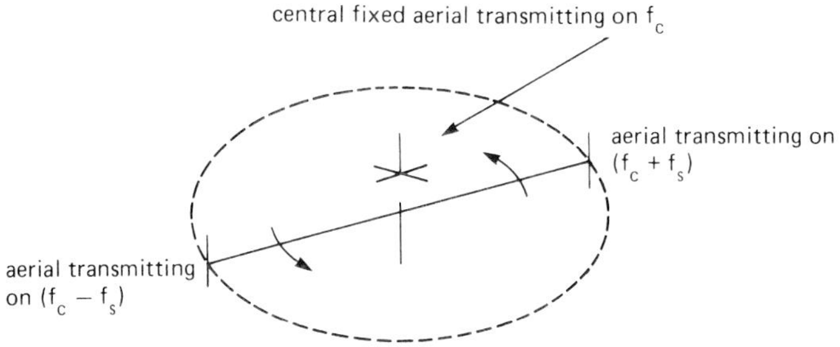

Fig. 21 Principle of double side-band Doppler VOR. The central aerial radiates the carrier frequency, f_c, one of the rotating aerials radiates f_c+f_s and the other f_c-f_s, where f_s is the frequency of the sub-carrier

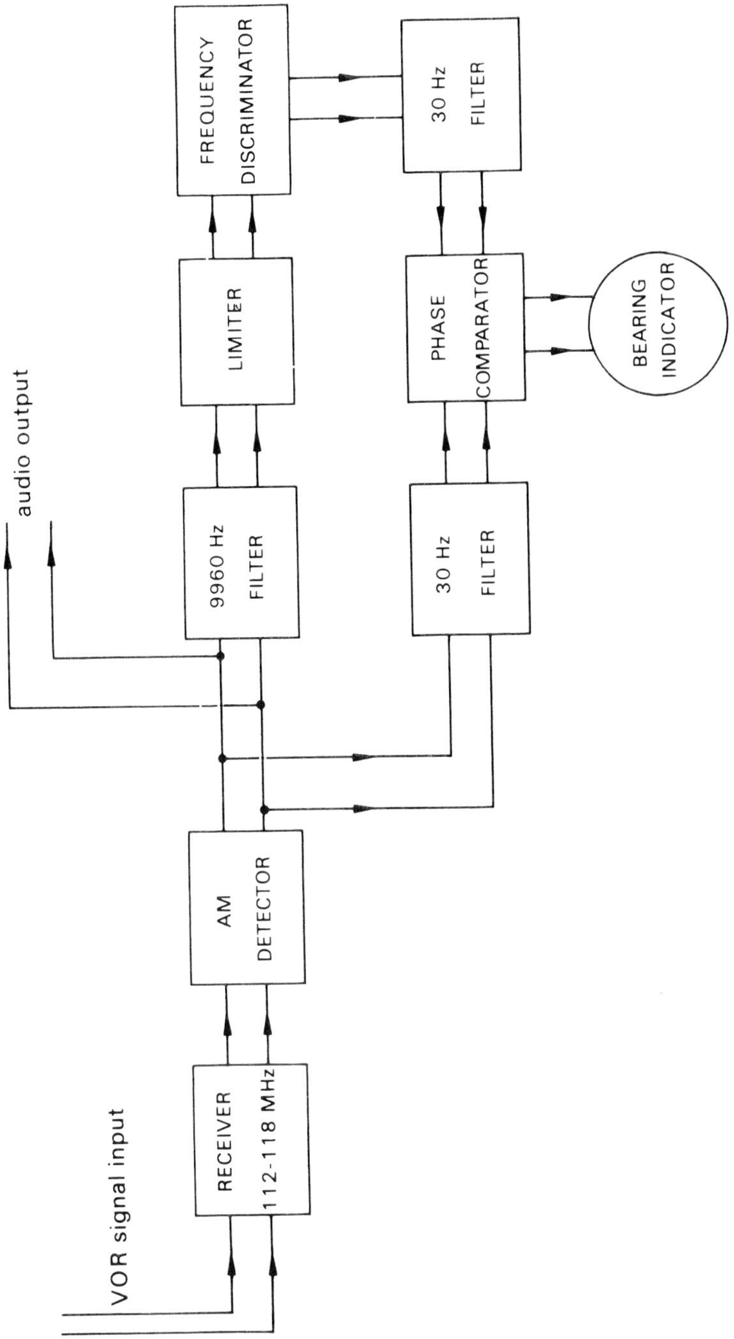

Fig. 22 Airborne VOR receiver

The airborne receiver

The airborne equipment consists of standard multichannel AM receivers in which the signal is split three ways after signal detection.

The first is connected to the aircraft intercommunication system to enable aircrew to identify the beacon being received and hear any voice transmissions radiated from the beacon.

The second and third pass through filters to separate the reference (9960 Hz) and variable (30 Hz) phase modulation. The reference phase signal is applied to a discriminator to recover the 30 Hz component and comparison is then made between the two 30 Hz signals to derive the bearing information.

Two types of display of VOR information are common, the first of which is called a Radio Magnetic Indicator and indicates directly relative bearing between the aircraft and the VOR station. This meter frequently has two indicating pointers which may be operated in conjunction with two VORs, two radio compass receivers or one of each. The second method of display uses the vertical pointer of an ILS meter. This operates in conjunction with a digital phase shifter on which any bearing may be preset. The ILS meter then indicates deviation from the pre-set bearing, allowing a VOR radial to be flown using a 'fly left/fly right' indication. The meter sensitivity is usually $\pm 10°$ for full scale deflection. To allow for whether the aircraft is flying towards or away from the beacon a sense-reversing 'to/from' switch is fitted.

Further output signals may be fed to the aircraft flight director system and for automatic updating of airborne navigation systems such as INS and Doppler.

4.3 Distance measuring equipment (DME)

Whilst VOR provides the pilot with an accurate determination of his *bearing* from a ground station, a means of measuring his *distance* from that station is necessary for him to ascertain his position. The equipment providing this information is Distance Measuring Equipment (DME) which operates on the secondary radar principle and is normally co-sited with the VOR station. The resulting combination forms the standard ICAO approved rho-theta short range navigation system.

Early DME equipments were developed from the wartime Rebecca-Eureka system which operated in the 200 MHz band. In 1946 agreement was reached to operate this service in the 1000 MHz band although it was not until some thirteen years later that the exact frequencies and pulse techniques were finally agreed.

The operation of the equipment is on the interrogator/transponder principle, the process being initiated by the aircraft.

The aircraft interrogation comprises a series of pulses, each 3.5 microseconds wide, radiated in pairs spaced by either 12 or 36 microseconds at a rate of between 5 and 150 pulse-pairs per second. On receiving a pulse-pair the ground transponder delays for 50 microseconds and then radiates a pulse pair of either 12 or 30 microseconds spacing on a frequency 63 MHz removed from the interrogator frequency.

The airborne equipment measures the time elapsed between initiating the interrogation and receipt of reply, subtracts the 50 microsecond beacon delay and displays the remainder as a distance presentation calibrated in nautical miles.

In such a relatively simple system the problem occurs of identifying the correct reply when the beacon is being interrogated by more than one aircraft. This difficulty is overcome by making the pulse repetition frequency of the airborne equipment somewhat unstable such that it will vary (within certain limits) in a random manner. The receiving equipment is designed so that only transmissions exactly corresponding in pulse repetition frequency to those radiated by the transponder are recognised and processed.

DME operating frequencies

Distance Measuring Equipment operates within the 960 MHz to 1215 MHz band. The interrogation and reply frequencies are allocated with 1 MHz spacing between adjacent channels and are numbered 1 to 126 in ascending order of airborne interrogating frequency from 1025 MHz to 1150 MHz. For

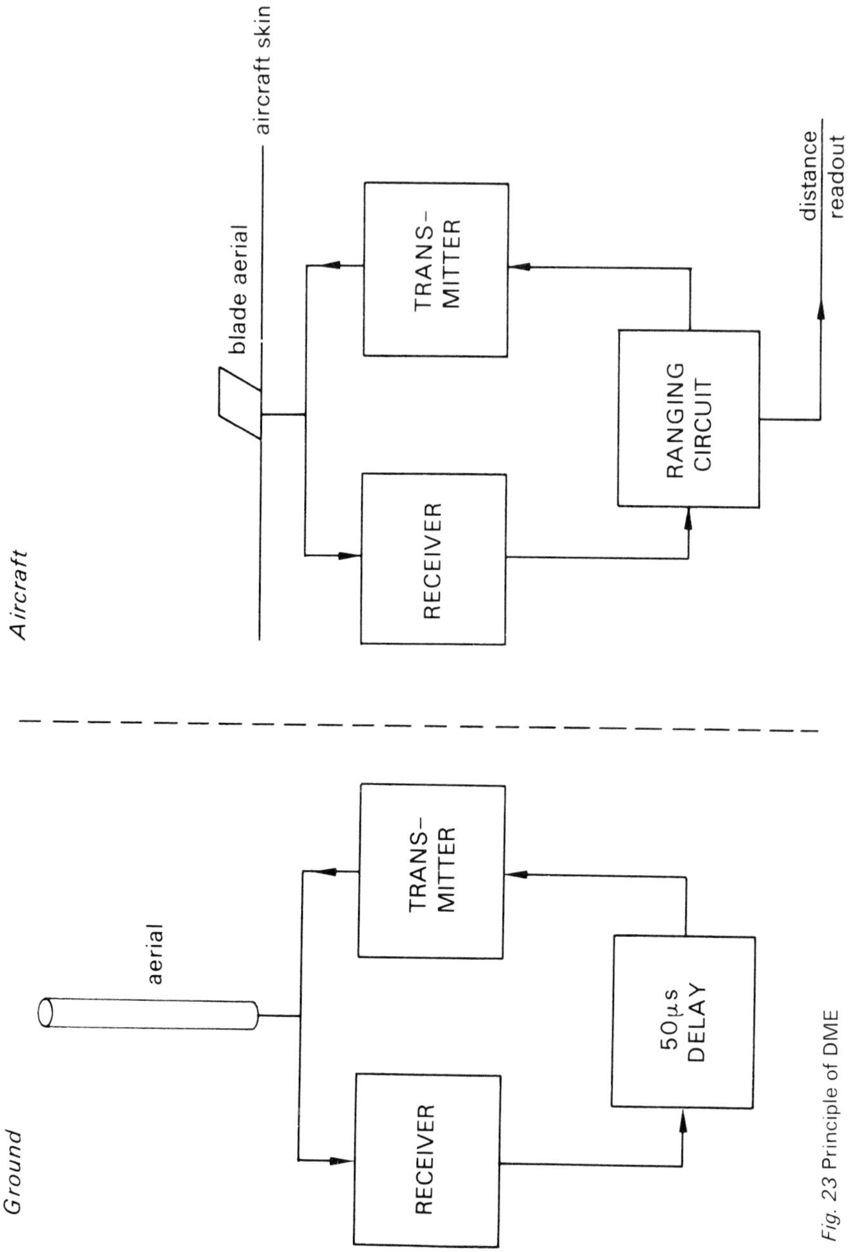

Fig. 23 Principle of DME

each airborne interrogating frequency two reply frequencies are allocated, one 63 MHz higher and the other 63 MHz lower in frequency. These are designated X and Y and vary not only in frequency but also in the pulse spacing of both interrogation and reply. X channels utilise a 12 microsecond spacing for both interrogation and reply whilst on Y channels a 36 microsecond spacing is used for interrogation with the beacon replying with pulses spaced 30 microseconds.

When DME is working in conjunction with VOR to form a single facility, its frequency of operation is determined by the frequency of the VOR beacon in accordance with ICAO recommendations. Thus a VOR on 112.30 MHz is always paired with a DME on Channel 70X (1094 MHz interrogate, 1157 MHz reply) whilst a VOR on 112.35 MHz would pair with Channel 70Y (1094 MHz interrogate, 1031 MHz reply). Channels 70 to 126 are allocated for this purpose.

DME may also be paired with ILS in order to give the landing aircraft a continuous 'distance to run' readout. In these circumstances, the 50 microsecond delay is modified such that a zero distance indication is received by the aircraft as it reaches the instrument runway touch-down point.

Airborne DME equipment

As the DME beacon reply is always 63 MHz displaced from the interrogator transmission, it is possible, by use of a 63 MHz receiver IF, to use a common oscillator chain for both transmit and receive sections of the airborne equipment. In earlier equipments either individual crystals for each of the 126 allocated frequencies or a crystal mixing process were used in the frequency determining circuits but in recent years these have been replaced by frequency synthesization techniques (see section 3). The output of the synthesizer or crystal multipler chain is fed either directly to the receiver mixer circuit or amplified further to the level necessary to drive the transmitter power amplifier.

It is evident that the receiver section of the interrogator will receive all transmissions from the DME beacon to which it is tuned and consequently circuits must be added to recognise authentic replies to interrogations. These are known as the 'searching' circuits. The principle of their operation is that a time 'gate' is generated, typically twenty microseconds wide, which sweeps slowly from a reply time corresponding to a distance of zero miles from the beacon to a time corresponding to the maximum range of the equipment, the output of the 'gate' being fed to an integrating circuit. This may take from one to twenty seconds and during this period the interrogator PRF is increased considerably. When the interrogator detects coincident replies from a series of consecutive pulses, it recognises these as replying to the aircraft's own transmission and switches the equipment from searching to tracking operation. In this mode the PRF is reduced to a minimum (five to twenty-five pulse-pairs per second) and additional circuitry is brought into service which ensures that the 'gate' tracks the received pulses. If the replies fall within the early part of the 'gate', it advances; if in the later part, the 'gate' is delayed.

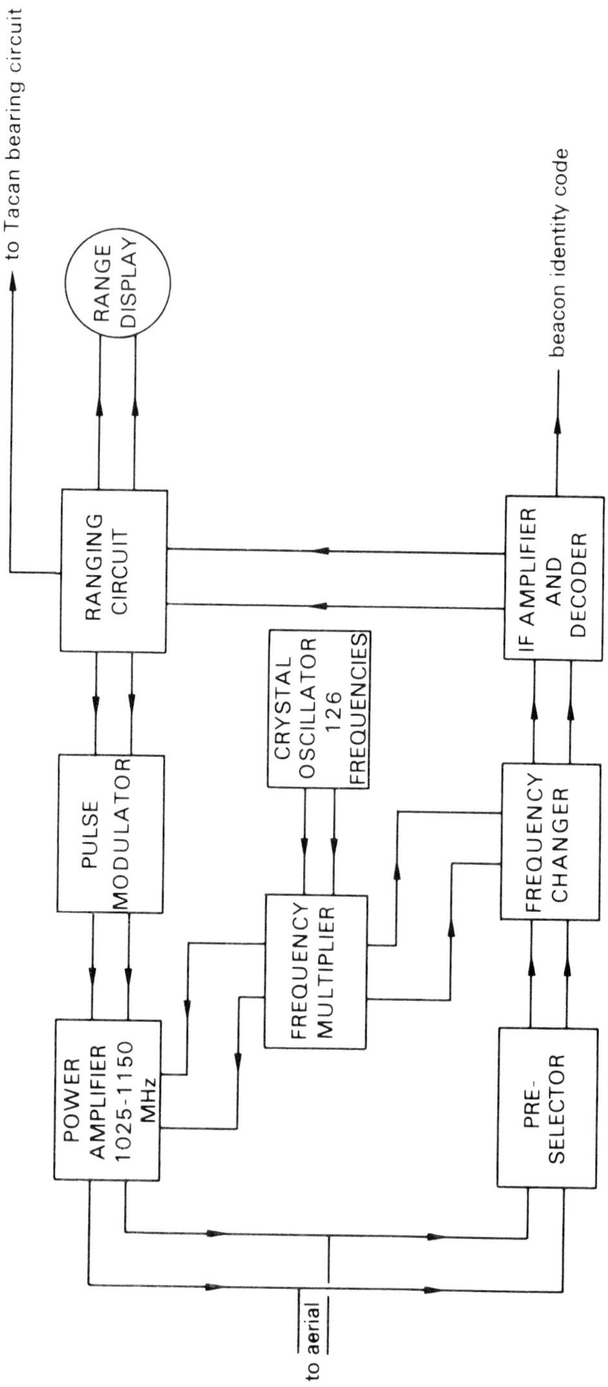

Fig. 24 DME airborne receiver-block schematic

Should the incoming signal be lost for some reason, the 'gate' remains static for a few seconds and if signal recovery is not effected, the equipment reverts to search mode. Alternatively the equipment may be designed such that the 'gate' continues moving at its last rate until re-acquisition of signal or reversion to search mode.

In the simpler, older equipment the position of the 'gate'–and consequently the range of the beacon–is displayed on an analogue meter calibrated in nautical miles. In more elaborate equipment either digital electro-mechanical displays, or more recently digital electronic displays are used. These can indicate to an accuracy limited only by that of the overall system.

DME ground equipment

As in the case of other ground navigational aids, the DME beacon is freed of many of the constraints of the airborne equipment as only one channel is in use at any one station and there are few power or space limitations. It is therefore possible to incorporate both more powerful transmitters and more sensitive receivers. Furthermore, lack of space limitation facilitates the use of high gain aerials, 9 dB being typical.

In the design of a ground DME beacon it is generally assumed that 95% of aircraft making use of the beacon will be in tracking mode, i.e. not exceeding twenty-five interrogations per second, and that the remaining 5% will be in the search mode at an interrogation frequency not exceeding 150 per second. Thus for 100 aircraft using the beacon simultaneously approximately 3000 pulse-pairs will be radiated. Most modern beacons operate on the constant duty-cycle principle and means are sought to maintain the output at about 3000 pulse-pairs per second.

This is achieved by increasing the receiver gain until the transmitter is being triggered by random atmospheric and receiver noise to offset any shortage of aircraft interrogations. In the absence of interrogation the receiver gain is increased to the point at which the required 3000 pulse-pairs are being triggered from noise alone, but when interrogations from less than 100 aircraft are being received the output is the result of a mixture of both noise and interrogations. If an excess of interrogations is received the receiver gain is reduced to limit the number of replies to the predetermined level.

The use of this system gives advantages in that the transmitter duty cycle remains reasonably constant, the beacon always remains in its most sensitive possible condition and in the case of excessive interrogation, the nearest aircraft are the last to lose service.

An extreme case of this would occur if the ground beacon were to fail for a brief period whilst being interrogated by 100 aircraft. Under such circumstances all the aircraft's equipment would go into 'search' mode increasing their PRF to 150 pulse-pairs per second. As the ground beacon returned to service it would be confronted by 15 000 pulse-pair interrogations per second. The ground receiver gain would then reduce until only 3000 interrogations per second were being received, this providing resumption of service to the nearest twenty aircraft. As track mode was restored in these receivers so their PRFs would revert to twenty-four or less pulse-pairs per second, allowing the ground receiver to increase in sensitivity to accept further, more distant aircraft in search mode. Gradually, after a few minutes, all aircraft would again be in the track condition with a total of less than 3000 interrogations per second.

Co-channel interference

When co-channel interference exists a low duty-cycle pulse aid such as DME has certain advantages. Signals on the same frequency will interleave but not add. Only the strongest of the stations have any effect on the receiver, the weaker ones being almost totally rejected. In contrast, CW systems, such as VOR, suffer from even a relatively weak interfering signal, which, although not taking control, may introduced large bearing errors.

System accuracy

A major factor in the accuracy of a DME beacon is the accuracy to which the 50 microsecond delay between reception of interrogation and transmission of pulse may be maintained and considerable effort is expended in maintaining this figure constant and independent of any external factors. The ICAO requirement states that overall system accuracy should be better than 0.5 mile or 3%, whichever is the greater, but of course, this must include all factors of which the accuracy of the delay is only part.

Identification

As with all navigational aids a morse code identification signal must be radiated at intervals. Typically this occurs at half minute intervals and is initiated by an external keying circuit usually common with the associated VOR. During the identification period the random pulses are replaced by regularly spaced pulses at a PRF of 1350 pulse-pairs per minute keyed with the beacon identification letters.

Plate 18 The Racal RDB100B watt non-directional beacon. Two transmitters are included, one of which is in standby mode but automatically takes over service in the event of main equipment failure.

Plate 19 A conventional VOR installation surmounted by a DME antenna. This equipment is situated near Nadi Airport, Fiji.

Plate 20 The Racal Mk2a DVOR *(Photo: Racal)*

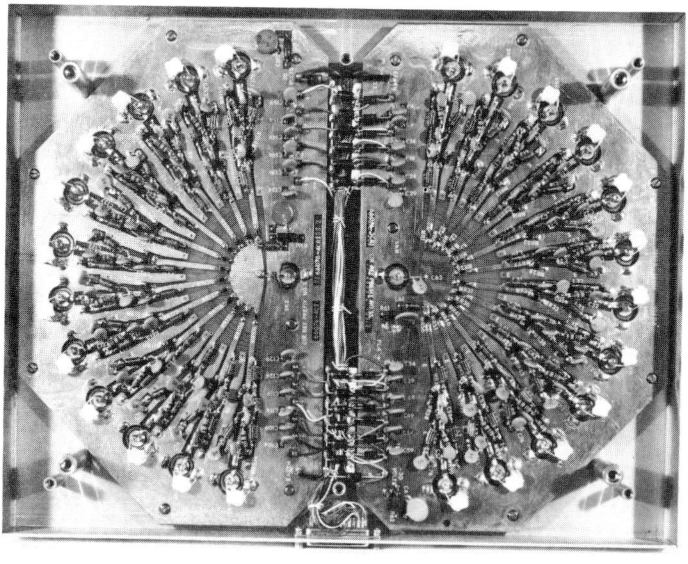

Plate 21 The fully solid state aerial switch used in the Racal Avionics DVOR equipment *(Photo: Racal Avionics)*

Plate 22 Concorde passes over the Racal Avionics localiser at Luton Airport. *(Photo: Racal Avionics)*

Plate 23 The glide-path aerials and cabin of the Racal fully solid state ILS system. *(Photo: Racal Avionics)*

Plate 24 Micronav MLS ground stations. Azimuth station on the right, elevation (glideslope) station on the left.

Plate 25 Another view of part of a Racal Avionics ILS Localiser antenna array. This one is installed at East Midlands Airport. *(Photo: Racal Avionics)*

4.4 Instrument landing system (ILS)

Instrument Landing System (ILS) is the standard approach aid in use throughout the world today. It is capable, under certain circumstances, of providing guidance data of such integrity that fully coupled approaches and landings may be achieved.

Introduced in 1946 in the form of the American SCS 51 equipment, succeeding generations of equipment showed improved performance until 1971, when, after incorporating the latest developments in solid state circuitry, the integrity of the system was deemed to be sufficiently high for categorisation for coupled approaches and landings by appropriately equipped aircraft.

Principles of operation

The system comprises three distinct equipments: the localiser transmitter, which gives guidance in the horizontal plane, the glide path transmitter which supplies vertical guidance and the two or three marker beacons, situated on the approach line which give an indication of 'distance to run' to the approaching aircraft. Each equipment includes automatic monitoring and remote control facilities.

The localiser is the lateral guidance portion of ILS and is situated at the up-wind end with its aerial on the centre line of the runway. Operating on a frequency in the 108.0 MHz to 112 MHz frequency band, it is required to radiate a signal modulated by 90 Hz and 150 Hz tones in which the 90 Hz predominates to the left hand (to the aircraft) of the approach path, and 150 Hz to the right. On the course line the modulation depth of each tone should be equal and on either side of this line the difference in depth of modulation (ddm) between the tones should be proportional to angular displacement. At 2° off course the ddm must be 15.5%, this corresponding to a 350 ft lateral displacement at the end of a 10 000 ft runway and also full scale deflection on the aircraft's ILS indicator. Outside this, to a limit of ±35° the ddm should be in excess of 15.5% to ensure that any approaching aircraft receives a full fly-left or fly-right indication. The guidance extends to a distance of 25 nm in the direction of the course line. This transmitter also emits an identification signal consisting of two or three morse coded letters.

The glide path (i.e. elevation guidance) equipment is situated at the side of the runway approximately adjacent to the touch-down point. The signal emitted, whose frequency is in the 328.6 MHz to 335.0 MHz band, again uses

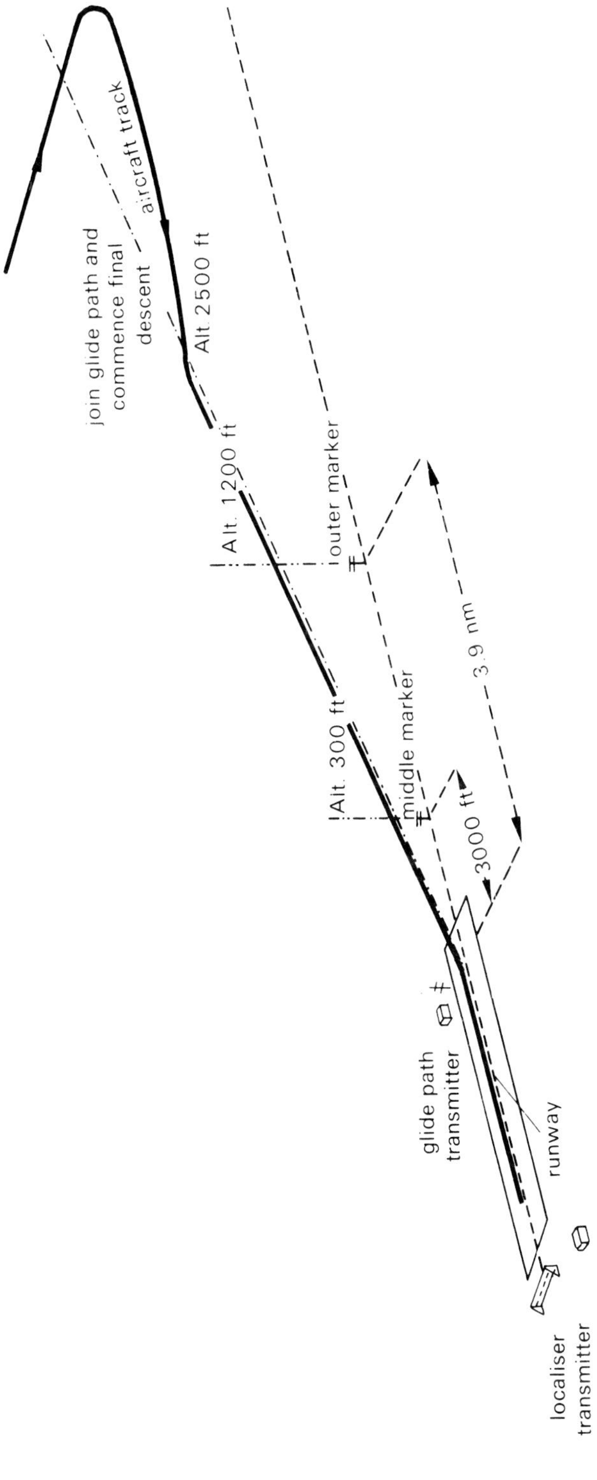

Fig. 25 ILS-system layout and approach path

90 Hz and 150 Hz tones, the 90 Hz predominating above the course line and 150 Hz below.

The signal defines a straight line approach path at an angle of between 2.5° and 3.5° above the horizontal to a range of at least 10 nm.

The system is completed by two or three marker beacons operating on 75 MHz. These are situated on the extended centre line of the runway approximately at 3.9 nm, 3000 ft and, optionally, between 1500 ft and 250 ft from the threshold of the runway and radiate a vertical 'fan', at right angles to the line of approach. Each beacon is distinctively coded and the radiation pattern ensures that each is heard for only a few seconds as the aircraft passes overhead.

Aerial systems

It is normal to achieve the localiser pattern by combining two separate transmissions, known as the course and clearance signals. These may be radiated from two equipments with individual aerials operating on closely adjacent frequencies or from one transmitter arranged to give two output signals of differing phase, one of which is fed to the whole aerial to generate the relatively sharp course signal and the other to only part of the array to radiate the broad clearance signal. The effect of the clearance signal is such that not only is the coverage extended but also any side-lobe effects from the main course array are also effectively masked.

To enable the course and clearance aerial arrays to radiate the required signal patterns, each has to be fed with two separate signals. The first, known as CSB (Combined Carrier and Sideband), consists of a normal A3 transmission, the carrier being modulated by both 90 Hz and 150 Hz signals. The second is the residue after the carrier of the modulated transmission has been phased out in an RF bridge, i.e. 90 Hz and 150 Hz sidebands only. This is therefore known as the SBO (Sidebands Only) output.

The localiser aerial array consists of a number (usually 12 or 24) of dipole or yagi aerials backed by a plane or horn reflector mounted in a line at right angles to the line of the runway. Each aerial is fed with CSB and SBO signals in varying proportions and phase such that, by the consequent combination of component signals, a narrow beam of the desired characteristic is radiated. The clearance signal is formed similarly either in a separate aerial array or in part of the main course array.

Due to the relatively shallow approach angle of incoming aircraft, the main lobe of the transmission must of necessity subtend only a small angle to the horizontal, thus it is inevitable that the surrounding ground is also illuminated. Any uneven geographical features such as hills, buildings, etc. underneath or adjacent to the flight path can cause a reflection of the signal which is receivable by the approaching aircraft.

Within the aircraft receiver the incoming signals combine, the resultant being a vector sum of the direct and all reflected signals, giving erroneous course indication, i.e. a beam bend. Although in practice beam bends cannot be entirely eliminated, current equipment has sufficiently sharp course beams to limit the area of ground illuminated and maintain such signal strength on the direct path transmission that consequent beam bends at most sites are of manageable proportions. Although on most difficult sites it has been found that the larger aerial arrays will provide guidance of sufficient integrity for most commercial purposes, there do remain, however, certain airports in very mountainous areas where it has not proved possible to operate a satisfactory ILS system.

The localiser transmitter

This transmitter is required to provide two stable transmissions on a single frequency in the 108.0 MHz to 112.0 MHz band, one modulated by 90 Hz and 150 Hz and the other, a double sideband suppressed carrier signal, the sidebands being in consequence of the 90 Hz and 150 Hz modulation.

The radio frequency signal is generated by a continuous wave transmitter of standard design which is then modulated by mechanical or electronic means.

The original SCS51 equipment used a mechanical modulation system, but during the 1950s, in an attempt to circumvent patents on this, several firms produced equipment using electronic modulation techniques, with varying degrees of success.

Through the 1960s and 70s, mechanical systems held the ascendancy, but by the early 1980s, the electronic system was again evaluated and by using solid state techniques, a level of reliability and stability was achieved which was unattainable by either the mechanical or earlier electronic systems.

As the operational life of an ILS installation is normally in the order of 15 to 20 years, systems using mechanical techniques will be in operation for many years to come, and thus it is appropriate to describe the principles of both.

The principle of mechanical modulation is based on that used in the 1930s two course visual aural range. The RF signal is divided two ways by an RF bridge. Adjacent to each of the output legs is a quarter wave section which absorbs power from that leg. A toothed wheel rotates within this section, alternately tuning and detuning and thus varying the amount of power absorbed from each output leg of the bridge. If the two toothed wheels, normally called paddles, have numbers of teeth in the ratio of 5:3, when rotated at the correct speed, the result will be modulations of 90 Hz and 150 Hz appearing on the output legs of the bridge. The two signals are recombined in a further RF bridge. An RF bridge may also be used to balance out the carrier from a CSB signal leaving SBO output.

In more modern designs the mechanical modulators operate within the RF bridges but the basic principle of operation remains the same.

The SBO and CSB RF outputs are then fed to the aerial distribution unit which is mounted adjacent to the aerial system and distributes the RF power to the individual elements to the aerial array.

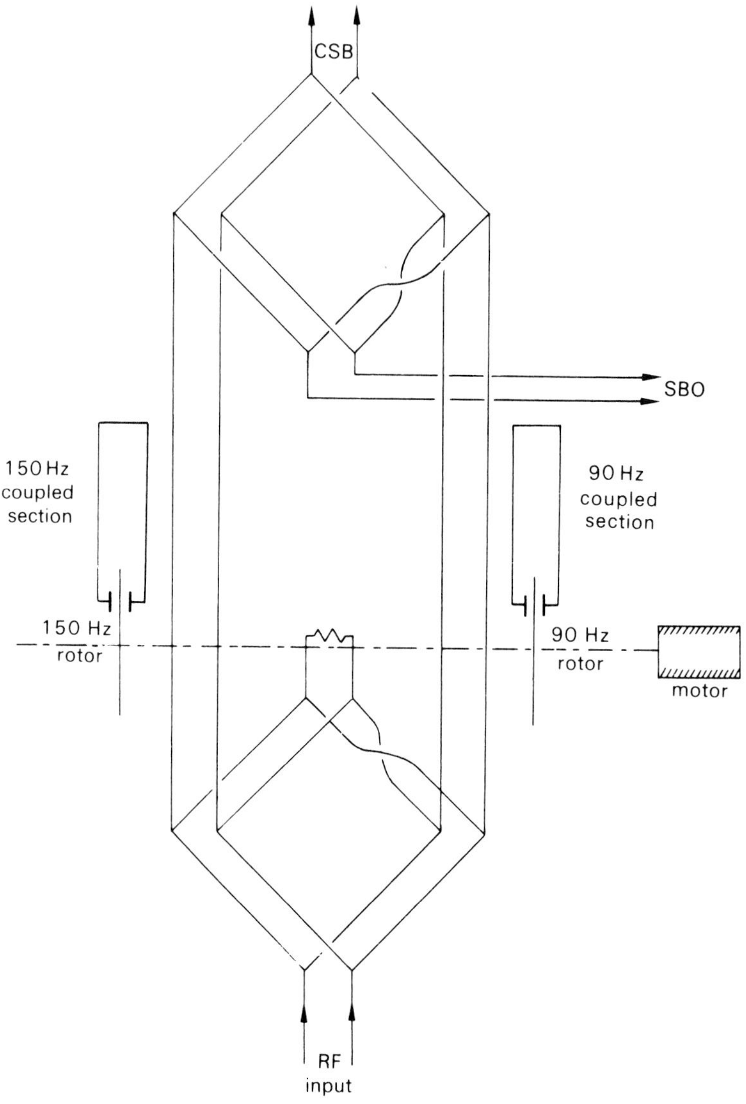

Fig. 26 Principle of mechanical modulator

In the electronic system of modulation, the output from the RF driver stages is divided into two paths. The first of these is amplitude modulated by the 90 and 150 Hz tones as in a standard communications transmitter. This is the CSB component. The second path is also modulated by the 90 and 150 Hz tones but this time in a balanced modulator circuit which also balances out the carrier wave, leaving only the sidebands, i.e. the SBO signal. Thereafter the CSB and SBO signals are fed to the aerials in the same way as for the mechanical modulation systems.

In mechanical modulation systems, the 90 and 150 Hz tones are inherently phase locked by their means of generation, but with the electronic system this is typically achieved by deriving each from a 450 Hz source and dividing by 5 and 3 respectively.

An interesting development has recently been introduced by the German manufacturer Standard Elektrik Lorenz who, rather than generating the required waveforms from independent 90 and 150 Hz signals, have digitally recorded the combined waveforms in binary code in random access memories. These are read and converted to analogue signals by digital to analogue converters, the output of which is then fed to the modulated stages of the transmitter.

The glide path

The earliest attempts to supply glide path guidance made use of a signal strength meter fitted to the aircraft receiver and the pilot followed a contour of equal field strength. This was not altogether satisfactory as it was realised that a straight line glide path was required. One of the earliest successes was the glide path of the US SCS 51 equipment. This equipment used two horizontal aerials, the lower at a height of $1\frac{1}{2}$ wavelengths and the upper at about 7 wavelengths above ground. The lower aerial was fed with carrier modulated by 90 Hz and the upper with carrier modulated by 150 Hz. It is a characteristic of the vertical polar diagram of a horizontal aerial that it will exhibit a number of lobes, the number being equal to the height of that aerial above ground measured in half-wavelengths at the operating frequency.

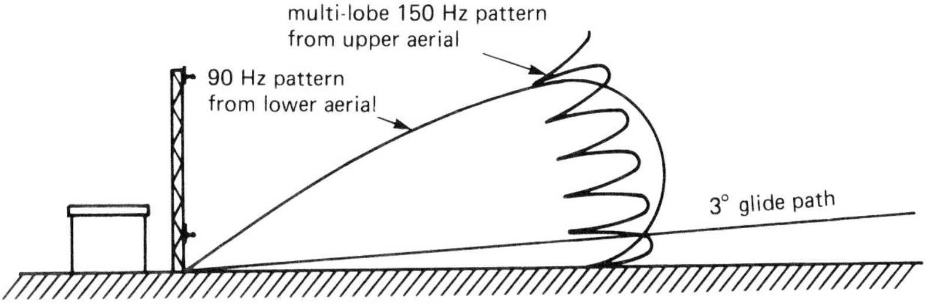

Fig. 27 Principle of equi-signal glide path

Considering therefore the lower aerial, three lobes will be transmitted, the lowest at about 10° above the horizontal. The upper aerial will radiate fourteen lobes with the lowest of these at a more acute angle to the ground than the lower lobe of the bottom aerial. At the intersection of these two lobes occurs the condition for ILS glide path guidance, i.e. a straight line path, on which the 90 Hz and 150 Hz modulation depths are equal, the 90 Hz modulation being above the desired approach line. The condition re-occurs at a much higher angle, causing what is known as a 'false glide path' but this is readily recognised by the aircraft due to the high descent rate necessary. Equi-signal glide path is the term applied to equipment operating on this principle.

This system is now all but obsolete, having been succeeded by more advanced systems but the description serves as a suitable introduction for all use the same general principle–interaction between the lobe structures of two or more aerials. Beam bends may be caused by ground reflections in a similar way to those affecting localiser transmissions. Several types of glide path aerial equipment are available, the aerial being selected to suit the site characteristics.

Three types of glide path aerials currently in use are:

(a) The null reference

This is the simplest of the three systems and is used on relatively unobstructed sites. SBO is radiated from the upper aerial and CSB from the lower.

(b) The M-array

This array is characterised by a third aerial. The complex signal pattern is less susceptible to re-radiated ground reflection signals due to an absence of radiation below 0.8° elevation.

(c) The sideband reference

This system is particularly suited to sites where the terrain falls away in the vicinity of the airfield. This type of aerial may be easily identified by the situation of the aerials on the mast, the upper of which is about one and a half times the height of the lower, both being significantly higher than those of the null reference system.

All three types of aerials are mounted to the side of the runway, the distance being related to obstacle clearance limits. To ensure that the guidance signals appear to originate from the runway centre line the aerials are 'stepped' towards the runway such that the upper aerial is closer to the runway than the lower though the distances involved are only a few inches.

The marker beacons

The specification ILS calls for a minimum of two marker beacons on the approach path. A third marker beacon may be added whenever, in the opinion of the Air Traffic Control Authority, an additional beacon is required because of operational procedures at a particular site.

The markers are positioned nominally at 3.9 miles, 3500 ft, and optionally between 250 ft and 1500 ft from the threshold of the runway. All operate on the same frequency (75 MHz) but each may be identified by modulation frequency and coding as follows:

marker	coding	modulation frequency
inner (where installed)	six dots per second continuously	3000 Hz
middle	alternate dots and dashes	1300 Hz
outer	two dashes per second continuously	400 Hz

The middle and outer marker beacons radiated a vertical 'fan' beam across the approach path, their location being selected such that at the outer marker height, distance and aircraft equipment functional checks may be made. The middle marker is located so as to indicate the imminence of visual guidance and the fact that the aircraft has reached category 1 decision height. The inner marker, where fitted, indicates that the aircraft is passing through category 2 decision height.

DME associated with ILS

As was stated in Chapter 4.3, DME is frequently paired with ILS in order to give the approaching pilot a distance to run indication. This is particularly useful when the approach path crosses an estuary or the open sea where it would otherwise be impossible to install the appropriate 75 MHz markers. In this service, the 50 microsecond DME delay is modified so that zero distance indication is given as the aircraft reaches the runway touchdown point.

Airborne equipment

The airborne equipment consists essentially of separate receivers to receive the marker, localiser and glide path guidance signals.

The marker receiver is a simple receiver tuned to 75 MHz the output from which is fed to both the aircraft intercommunication system and to an indicator lamp on the pilot's instrument panel.

The localiser and glide path receivers are conventional crystal controlled amplitude modulation equipment with their frequency selectors 'ganged'. This is possible as ILS localiser and glide path frequencies are 'paired'. After detection within each receiver the signals are fed to filters to separate the 90 Hz and 150 Hz components which are compared and used to drive the horizontal and vertical indicating needles on a cross pointer meter. Also included within the ILS cross pointer meter are two 'flag' indicators. These are energised from a circuit which summates the tone levels and thus indicates that guidance information is present. An additional unfiltered output is taken from the localiser receiver to the aircraft intercommunication system so that the crew may hear the station identification signals radiated from the localiser transmitter.

The sensitivity of the indicating instrument is arranged so that 15.5% ddm causes full scale deflection of each needle. The localiser receiver energises the vertical needle, the deflection being to the right when 90 Hz predominates, i.e. the aircraft is to left of course. The horizontal needle is controlled by the glide path receiver and moves downwards when 90 Hz predominates. It will therefore be seen that for a pilot to regain correct course alignment it is necessary to 'follow' the meter indication. The ILS output signals may also be coupled to the aircraft flight director system to facilitate coupled approach and landing procedures. In these circumstances it is usual to operate two or three equipments simultaneously, the flight director taking instructions when both signals are in agreement for a duplicated system, or a minimum of 2:1 majority in the case of triplicated systems.

Categorisation of ILS

ILS is categorised in two ways–'Operational' and 'Facilities'. The operational categories are couched in general terms and are recommended minima, while the facility categories are equipment specifications recommended by ICAO. These, in conjunction with a satisfactory flight inspection will enable the ILS to be certified to give an airfield an operational category.

The operational and facilities categorisations are as follows:

Category 1

Operational	Operation down to 60 m decision height with Runway Visual Range (RVR) in excess of 800 m with high probability of success.
Facility	An ILS which provides guidance information from the coverage limit to the point at which the localiser course intersects the glide path at a height of 60 m above the horizontal plane containing the threshold of the runway.

Category 2

Operational	Operation down to 30 m decision height with RVR in excess of 400 m with high probability of success.
Facility	An ILS which provides guidance from the coverage limit to a point where the localiser course line intersects the glide path at a height of 15 m or less above threshold level.

Category 3

Operational 3a Operation with no height limitation to and along the surface of the runway with external visual reference during the final phase of landing with RVR of 200 m.

Further objectives are categories 3b and 3c with RVRs of 45 m and zero respectively. These require guidance along the runway and 3c guidance to the parking bay.

Facility An ILS which, with the aid of ancillary equipment if necessary, provides guidance from the coverage limit to, and along, the surface of the runway.

ILS definitions

Facility reliability

The reliability of an ILS is defined as the probability that a ground installation radiates signals within specified tolerances.

Integrity

This is that quality which relates to the trust which can be placed in the accuracy of the information supplied by the facility.

Reference datum

A point at a specified height located vertically above the intersection of the runway centre line and threshold and through which the extended straight portion of the ILS glide path passes.

4.5 Microwave landing systems (MLS)

Instrument Landing System (ILS) which is today the universal aid for approach and landing was originally standardised by the International Civil Aviation Organisation in 1949. After continuous development the present ILS equipment is adequate for the air traffic of today and in the reasonably near future. However, it became evident by the late 1960s that it was necessary to consider an eventual replacement and in 1972 ICAO published an operational requirement for a new non-visual approach and landing guidance system.

The justification for such a system is based on the shortcomings of ILS which will become more significant with future growth of air traffic and a need to maintain regularity and safety in all weather conditions. The principal drawbacks of ILS are:

(a) Approaches are confined to a single narrow path.
(b) The number of channels is limited.
(c) The quality of the guidance signals is dependent on the nature of the terrain and can, for example, be seriously affected by snowfall. In consequence, the siting of ILS at some airports can be both difficult and expensive and at a few airports, impossible.

Technological improvements since ILS was introduced have made the use of microwaves a practical proposition and at these frequencies significant independence from site conditions can be achieved. Aerial radiation patterns can be tailored to lift the radiation from the ground using arrays which, although large electrically, are relatively small in physical terms.

The ICAO preference

The All Weather Operations Panel (AWOP), which was set up by the Air Navigation Commission (ANC) of ICAO, was charged with evaluating alternative proposals for this new system. The panel met in the spring of 1977 and considered four alternative systems: Interscan (Australia) and Time Referenced Scanning Beam (USA), which are similar in principle, Doppler (UK) and DLS (Federal Republic of Germany); finally recommending the TRSB/Interscan proposals. This was adopted by the AWO divisional meeting in Montreal during April 1978.

In coming to this decision the meeting noted that 'regardless of some of its limitations that led to the development of a new guidance system, the present-day ILS represents a world-wide, well-established and reliable system, offering safe and efficient services within its technical and operational capabilities.' It further noted that 'large investments have been, and will be, made during the coming years in both ground and airborne equipment.' In order to allow sufficient time for amortisation of these investments it was agreed that the protection date for ILS should be 1 January, 1995 and at least until that date:

(a) ILS will remain as an ICAO standard non-visual aid for approach and landing.
(b) No state will be required to install MLS to provide service to aircraft.
(c) No air operator will be required to install the MLS because of withdrawal of ILS service at any international airport.

The time reference scanning beam/interscan system

This system is an air derived system in which ground based equipments transmit position information signals to a receiver in the landing aircraft. The angle measurements are derived by measuring the time difference between successive passes of highly directive narrow fan-shaped beams and distance measurements are obtained from a suitably located DME. All angle and auxiliary data functions are accommodated on the same assigned channel by a system of time division multiplexing. Thus a single receiver processor in the aircraft may sequentially process all of these signals. Each function is an independent entity within the time-multiplexed format, identified by a function preamble, detection of which sets up the receiver processing circuitry to decode the remainder of the function transmission. On completion of the decoding process, the receiver awaits the reception of the next function preamble whereupon the process is repeated. In addition to angle information, allowance is also made within the format for the inclusion of auxiliary data which may be transmitted using a function identification code and an identifying address preceding the data word itself.

The angle functions radiated within the TRSB/Interscan format are: azimuth, elevation, missed-approach azimuth, flare and 360° azimuth.

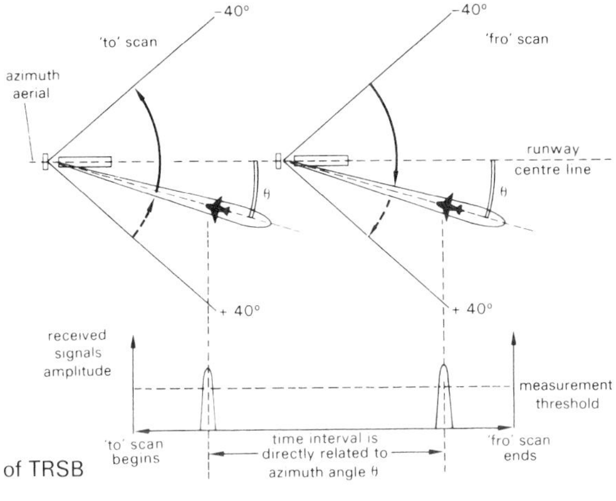

Fig. 28 Principle of TRSB

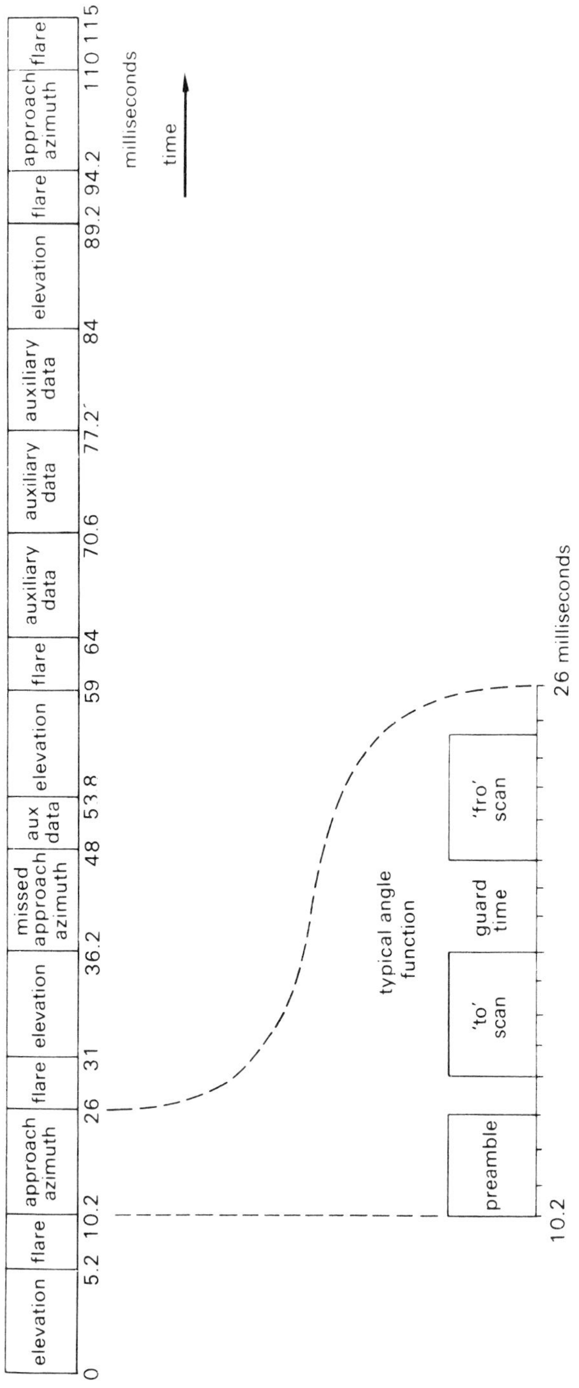

Fig. 29 TRSB time-division multiplex format

Angle guidance

Each angle function transmission consists of four elements:

(a) The preamble. This consists of a five unit synchronising code plus a function identity code of five information bits plus a parity bit.

(b) A series of pulses for left/right guidance and Out of Coverage Indication (OCI). The left/right pulses are transmitted in the function format to extend the TRSB coverage sector. The OCI pulses are transmitted by approach azimuth and missed approach azimuth systems, to provide proper flag indication when flying outside the system coverage sector.

(c) The 'to' and 'fro' angle scan.
In the azimuth guidance transmission the 'to' beam is scanned with uniform speed, starting from the clockwise extremity of the coverage limit (as viewed from above) and moving towards the other. After a short guard time, the beam then scans back to the starting point, thus producing the 'fro' scan for azimuth. For each scanning cycle, two pulses are received in the aircraft, the time interval between these being proportional to the angular position of the aircraft with respect to the runway.
The elevation function operates in a similar fashion with the beam first scanning upwards and then downwards. The TRSB format allows for 13.5 scans per second in azimuth and 40.5 scans per second in elevation.

(d) A pair of pulses for system test which may be used by receivers or an end to end check in receiver test mode.

Frequencies and channellisation

The time multiplexing technique allows all functions to take place on a single channel. A total of 200 channels have been allocated, spaced 300 kHz apart in the frequency band between 5031.0 MHz and 5090.7 MHz.

Auxiliary and basic data

In addition to positional information, the TRSB format provides for the transmission of data on two levels, i.e. basic and auxiliary. The basic data are designed for all system users and give information on basic system parameters such as minimum selectable glide slope, azimuth coverage limits etc. The auxiliary data are intended for the more highly equipped aircraft and give such information as runway conditions and siting data in a form which allows simple processing for presentation on standard displays. The auxiliary data format requires an airborne processor which is capable of handling sixty-four bit words.

Section 5

Radar

5.1 Primary ground radar

There are two types of ground radar–primary and secondary. Each relies for its operation on the transmission of a highly directional burst of radio energy which impinges on the target. In the case of primary radar, the signal reflected from the target aircraft is received and the time from transmission of the pulse to reception of the reflection is measured. The direction of the aerial indicates the bearing of the target aircraft and as the speed of radio waves is known, the time difference between transmission and reception gives the range.

With secondary radar, no use is made of the reflected signal, the pulse being received by the aircraft and used to trigger a transmitter on an adjacent frequency. This signal is received at the radar station and the range and direction of the target can be determined in a similar way to primary radar. Furthermore, the aircraft response to the interrogation can be coded to provide information such as identification and height.

Primary radar

The operation of a primary radar equipment is directly analogous to the classic example of a man measuring his distance from a cliff by shouting and using a stop watch to time the return of the echo. If that period was, say, four seconds, assuming the speed of sound is 330 m/s, the sound will have travelled $4 \times 330 = 1320$ m, i.e. 660 m in each direction, this being the distance between man and cliff.

In radar, as in the classic case, we again measure the transit time for the two way journey but as the velocity of radio waves is 300×10^6 m/s, the time intervals are so small it is more convenient to measure in millionths of a second or microseconds. Thus if the man mentioned previously had used radar, the time interval between transmission and reception would have been

$$\frac{1320}{300 \times 10^6} = 4.4 \text{ microseconds.}$$

By convention, within the aviation context, speeds are measured in knots (nautical miles per hour) and distance in nautical miles. It is convenient to remember that the time taken for a radar pulse to travel 1 nm and return is 12.36 microseconds.

The radar equation

It is obvious that in order that a radar range measurement may be made, sufficient power must be reflected from the target for a usable signal to be received at the radar site.

Many factors affect the maximum performance of radar equipment, but by using geometric principles it is possible to derive an equation which will give a measure of radar range.

First let us consider a radar transmitter of power P which radiates equally in all directions. The power density of the signal impinging on a target at range r will be:

$$\frac{P_t}{4 \pi r^2}$$

However, in real life, the transmitting aerial has a gain of G_t, so the power density at the target will be:

$$\frac{P_t G_t}{4 \pi r^2}$$

If the target area is S, then the power intercepted will be:

$$\frac{S P_t G_t}{4 \pi r^2}$$

For the sake of simplicity, let us assume that this power is re-radiated in all directions. The power density received from this back at the radar head will be:

$$\frac{S P_t G_t}{16 \pi^2 r^4}$$

Now the power entering the radar receiver will be related to the effectiveness of the aerial which, in turn, is related to its area and efficiency. This is usually referred to as its effective area and represented by A_o.

The power reaching the receiver (P_r) is therefore:

$$\frac{S A_o P_t G_t}{16 \pi^2 r^4}$$

Now A_o is related to the gain (G_r) of the receiving aerial and to the area of the aperture A. A further constant, k, allows for the difference in efficiency between an ideal and a practical aerial design, thus:

$$G_r = \frac{4 \pi}{\lambda^2} \times kA$$

Rearranging this equation and substituting in the previous equation we get:

$$P_r = \frac{S\, G_r\, \lambda^2\, G_t\, P_t}{64\, \pi^3\, r^4}$$

As the same aerial is used for both transmitting and receiving, we can rewrite the equation as:

$$P_r = \frac{S\, k^2\, A^2\, P_t}{4\, \pi\, \lambda^2\, r^4}$$

Up to this point we have assumed that the target re-radiates power equally in all directions. This, of course, does not happen in real life, so we now have to introduce the concept of 'equivalent echoing area'. This is the cross-sectional area which an isotropic radiator would require to return a signal similar to the actual target. It is usually denoted by σ. This is a theoretical figure, for the signal returned by the aircraft can depend on many factors, including shape, size, aspect, etc.

Rearranging the expression for received power and substituting σ for S we get:

$$r^4 = \frac{k^2\, A^2\, \sigma\, P_t}{4\, \pi\, \lambda^2\, P_r}$$

This equation gives the range for a given received power. For a specific radar equipment, the greatest range at which a given target can be detected will depend on the smallest value of P which will give a recognisable echo. If we call this $P_{r\,(min)}$ we can now write:

$$r_{(max)} = \sqrt[4]{\left(\frac{k^2\, A^2\, \sigma\, P_t}{4\, \pi\, \lambda^2\, P_{r(min)}}\right)}$$

Examination of this equation invites the conclusion that the shorter the wavelength, the greater the range. However, in this we are ignoring effects which changes of wavelength have on the effective echoing area of the target and on atmospheric attenuation.

Great care must therefore be taken in choosing the operating frequency of a radar to ensure that all factors are adequately considered.

Radar equipment in its most basic form

Basic radar equipment is required to emit a high power burst of radio energy in a narrow beam for a brief period of time–typically one microsecond–and then switch to a receiving mode. Any echoes received are amplified, processed and fed to the display equipment. This cycle of operation is repeated at the Pulse Recurrence Frequency (PRF). The equipment must therefore comprise: aerial, transmitter, receiver and signal processing equipment.

Under certain circumstances, however, when it is not possible to generate a pulse of sufficient power for the intended purpose, a much longer pulse may be transmitted and the returning echo compressed to give the effect of a short, very high power pulse. This technique is known as Pulse Compression and is described later in this chapter.

The purpose of the aerial is to radiate the output signal from transmitter in a sharp beam in the horizontal plane. In early VHF radar equipments the aerial array took the form of large numbers of dipole aerials spaced and fed such that the radiation from each dipole combined to form the requisite beam. With the transfer to shorter wave lengths, this method fell from favour but in recent years a more highly developed version of the system has been introduced and it is now finding favour particularly in airborne equipment. The most common type of aerial at present in use for surveillance equipment takes the form of a simple aerial illuminating a large reflector specially shaped to generate the required directional characteristics in the vertical and horizontal planes. As frequency increases so the efficiency of conventional coaxial feeder cables decrease to the point where, on many installations, the majority of the output power of the equipment would be absorbed between the transmitter and the aerial array. This has necessitated the use of waveguide feeders which comprise a metal–usually brass–rectangular section tube. The theory of operation of waveguides is very complex, but in general if the cross-sectional dimensions of the guide are approximately 0.7 × 0.4 wavelengths, it will act as an extremely efficient feeder system. Using this system, complexities arise due to its inflexibility but these are overcome by the use of special rotating joints and flexible sections. At signal frequencies the use of conventional feeders is limited to short interconnecting leads.

The radar transmitter consists of two circuits: the radio frequency generator and the modulator. The generator may take one of two forms, either a single oscillator such as a magnetron, or a multi-stage transmitter using a Klystron, Travelling Wave Tube or Multiple Transistor Module in the output stage. The output power generated depends on the ultimate purpose for which the equipment is intended. This may vary from 25 kW peak for a small twenty-five mile radar to several megawatts for long distance surveillance.

In section 3 (Radio Telephony Communication) the action of an AM modulator is described. However, in radar equipment the function is different. It is to switch the transmission on and off. Frequently the modulator actually supplies the necessary power to the transmitter for the required period of transmission. As no mechanical device could operate with the necessary stability at the required speed, the switching is entirely electronic.

Radar receivers have characteristics that set them apart from normal radio receivers. Firstly, the signals that they are required to receive may vary in strength in the range of one million to one between the strongest and the weakest. Secondly, the receiver has to have a very wide bandwidth in order that the echoes can be received with minimum distortion.

Even the best receivers inadvertently generate some internal noise and as the weaker echoes received may well only compare with this level it is of paramount importance that the internal noise generated within the receiver be maintained at the lowest possible level. It also follows that the ratio of signal to noise at the output of the receiver will be inferior to that of the input. A measure of the quality of a receiver is a comparison between input and output signal-to-noise ratios. This is known as the 'noise factor' and is expressed in decibels.

Early radar receivers used no pre-mixer amplification, achieving a noise factor in the order of 10–12 dB but by the early 1960s signal frequency amplification using parametric amplifiers and Travelling Wave Tubes (TWTs) facilitated an improvement to 3–5 dB. In recent years developments in solid state technology have enabled UHF transistor amplifiers to achieve noise factors better than 2 dB. Although the latter have considerable advantages in terms of power requirements and simplicity of circuit design, economic factors dictate that some of the older amplifiers will be in service for many years to come.

From the RF amplifier the incoming signal is fed to the mixer stage. For many years mixer circuits using waveguide technology reigned supreme but in modern equipment these have been replaced by printed circuit technology using microstrip techniques.

The local oscillator associated with the mixer stage in early years used a reflex Klystron valve but requirements of stability and ease of tuning soon led to the use of specially designed thermionic triode valves. Needless to say, in the most modern equipments thermionic valves have been replaced by UHF transistors.

The level of IF signal output from the mixer stage is still relatively low and some further amplification is desirable before transference to later stages of the equipment. This is achieved by the IF head amplifier which is situated adjacent to the mixer stage and whose purpose is to boost the level of the IF signal to such a level that any noise picked up on succeeding interconnecting leads will be insignificant compared with the signal and that any cable or interconnection losses will be more than adequately overcome.

It is within the main IF amplifier that a radar receiver varies most from its communications counterpart. Mathematical analysis shows that the transmission of a narrow pulse generates considerable sideband energy over a bandwidth of several megahertz, consequently, to receive such a transmission the bandwidth of the receiver must be equivalent as any reduction would degrade the pulse shape received and, in consequence, accuracy in ranging. The broadbanding arrangements used in the past considerably reduced the gain of individual amplifier stages so that many more stages were necessary than in non-radar receivers.

In common with almost all other branches of electronics the introduction of integrated circuits with their intrinsically high gain and stability has very much simplified the design of modern IF amplifiers enabling the necessary bandwidth and gain to be achieved in possibly only two or three stages.

Reference has already been made to the requirement for a radar receiver to function efficiently, receiving signals that vary in strength by a ratio of a million to one. The strongest returns are those from targets closest to the transmitter and such echoes are the earliest to be received after the radar pulse has been transmitted. It is therefore possible to vary the gain of the receiver during each reception period from a low value immediately after the emission of the pulse, increasing with time such that full sensitivity is achieved at a time corresponding to a range where no saturating echoes are liable to be received. The depth of control and the length of time over which it

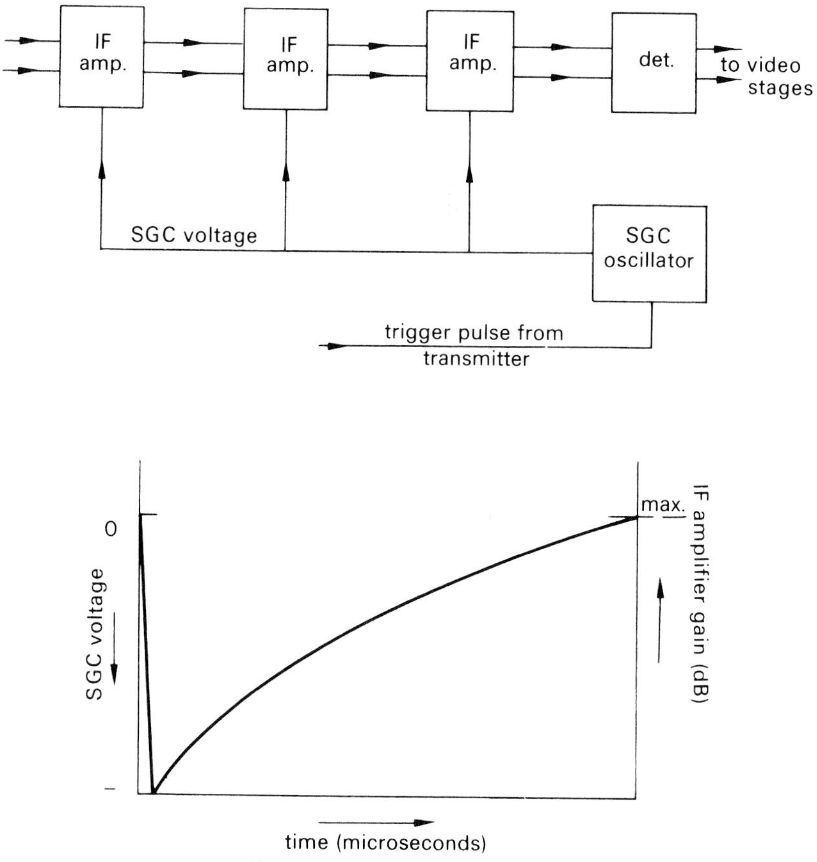

Fig. 30 Swept gain control

operates are variable and are normally determined by trial and error to suit local conditions at the time of commissioning the equipment. This gain function is known as either Sensitivity Time Control (STC) or Swept Gain Control (SGC), the latter expression being more common in the UK.

A further method of reducing saturation effects is by the use of a 'logarithmic receiver'. This is an alternative IF strip designed such that a detector is connected across the output of each IF stage, the outputs of the detectors being added together. If the input to such a receiver is increased, the last IF stage will be the first to saturate. After a further increase, the penultimate stage will saturate and so on until all stages reach saturation. This technique greatly extends the dynamic range of the receiver.

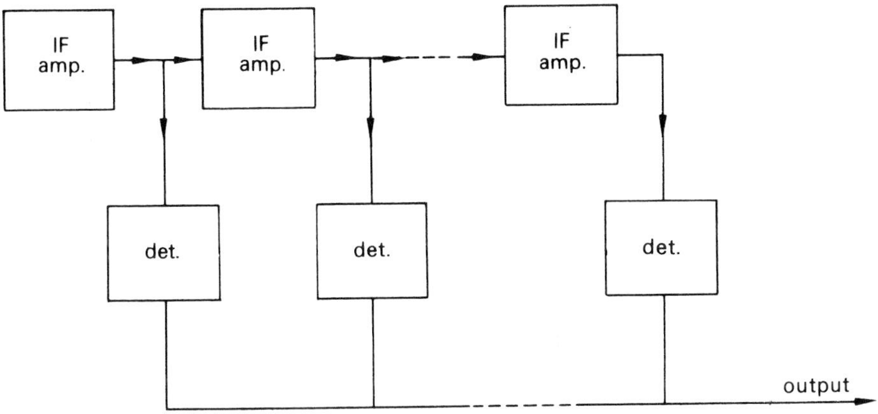

Fig. 31 Logarithmic receiver–schematic

The IF amplifier is succeeded by the detector. This is similar to a standard AM detector with the exception that component values are chosen to suit the wide bandwidth of modulation frequencies involved. The circuitry of the radar receiver is then completed by the video amplifier and processing stages.

After detection, the incoming signal can be fed directly to the display, but in many installations the displays may be a considerable distance from the receiver. Furthermore, the preferred input to displays is signals of constant peak level and signal to noise ratio. It is therefore usual to fit a stage of amplification after the detector and feed its output to 'limiter' circuits. These consist of a pair of electronic 'clipper' circuits, one limiting the peak output and the other eliminating all signals below a certain level. Both limiters are adjustable, the normal setting of the latter (known as the 'base clip') being such as to only reduce the noise level. By simultaneous adjustment of the two controls it is possible to accurately adjust the peak level and signal to noise ratio of the signals being fed to the display system.

Radar frequencies

Radio and radar wavebands are, for convenience, each identified by a letter. This practice originated in World War 2, and the identifications allocated at that time are still in common use. However, in the intervening years an alternative identification system with much wider application has been introduced.

The most commonly used radar bands (identified by the older designation with the more modern terminology in brackets), are:

L-Band (D): about 1000 MHz. Radar equipment in this band is normally used for long-range surveillance. Due to the low frequency, the aerials are large but the use of this frequency achieves good penetration through cloud and precipitation with consequently little interference from natural elements. Range is good and frequently exceeds 250 nm for high flying aircraft.

S-Band (E): about 3000 MHz (10 cm wavelength). This is the waveband usually chosen for airfield surveillance as it allows reasonable range of sixty to one hundred miles in conjunction with a sufficiently narrow beamwidth to permit Surveillance Radar Approach guidance (SRA) down to a range of two miles or so. Heavy precipitation can cause reflections but this effect may be minimised by the use of Circular Polarisation (CP).

X-Band (I): 10000 MHz (3 cm wavelength). This further increase in frequency enables even sharper beamwidths than on S-Band to be achieved. The precision elements of Precision Approach Radar (PAR) normally operate on these frequencies enabling aircraft approaches to be accurately controlled to a distance of a quarter of a mile from touch-down point. Some airports also have small surveillance radars operating in this band which are suitable for SRAs to closer limits (half a mile) than S-Band equipment.

Equipment operating on X-Band is very susceptible to precipitation returns thus making it suitable for airborne cloud/collision warning but the use of circular polarisation on ground equipment can minimise such interference.

Q-Band (K): 35000 MHz. As equipment operating on this band is highly susceptible to interference from precipitation it is used for short range applications only. However, due to the very sharp beamwidth attainable with aerials of only moderate size it has, in the past, been used for airport ground control. An example of this was the equipment which was formerly in use at London's Heathrow Airport, which was quite capable of showing a man standing on the airfield.

Modern developments have led to a decline in the use of Q-Band for this purpose, as modern signal processing techniques have enabled equivalent definition to be obtained from X-Band equipment with consequent financial and maintenance advantages. It therefore seems unlikely that further Q-Band equipment will be installed for Aerodrome Surveillance in the foreseeable future.

Pulse length and range discrimination

The most important consideration in the choice of pulse length is that of range discrimination. As previously described, a radio or radar wave travels 300 m in one microsecond, thus a pulse one microsecond in duration will extend 300 m along the direction of propagation. If two targets happen to be within that 300 m they will be illuminated simultaneously by the pulse and return only a single echo to the receiver. Conversely, the longer the pulse, the longer the target will be illuminated by the beam and the greater the chance of a good return being received.

The choice of pulse length is therefore a compromise between conflicting requirements and in practice a length of one microsecond is satisfactory for most surveillance radars.

PRF

The number of pulses transmitted each second by a radar transmitter is known as the Pulse Recurrence (or Repetition) Frequency or PRF. This frequency is governed by the fact that time must be allowed for each pulse to reach a target at the maximum range of the equipment and return. The transmitted pulse, however, does not stop at the maximum range, and may hit more distant targets. A further period is therefore allowed for any returns from targets beyond the maximum designed range of the equipment. This is known as the 'dead time'.

A typical airfield surveillance radar may have a range of 60 nm. The time for a pulse to reach a target at this maximum range and return is therefore 742 microseconds. However the PRF will normally be in the order of 450, i.e. 2222 microseconds spacing, allowing a dead time of 1480 microseconds. Notwithstanding the considerable allowance of 'dead time', under certain meteorological conditions anomolous radio propagation occurs and returns may be received from several hundred miles. Such echoes may be received during a succeeding reception time and take the form of an unknown bank of permanent echoes appearing anywhere on the screen. In general, the phenomenon is readily recognised by experienced staff and rarely causes other than slight inconvenience. However, in certain regions of the world, anomalous propagation is sufficiently common and severe to warrant special processing techniques to render the picture usable.

Aerial beam width and angular discrimination

The radar parameters so far discussed, have concerned the ability of the equipment to determine the range of the target aircraft. To enable the radar 'fix' to be made, the bearing from a predetermined point, i.e. the radar head, has also to be measured. This is achieved by arranging that the radar aerial emits only a narrow beam of energy in the horizontal plane so that the target will be illuminated only for the period that the aerial is pointing directly at the target.

As the direction to which the aerial is pointing can be determined to a considerable degree of accuracy, this, combined with range information, enables the aircraft position to be calculated.

The accuracy to which the bearing of the target may be determined is dependent on two factors:

(a) The width and symmetry of the beam transmitted.
(b) The accuracy to which the aerial position can be determined and relayed to the radar display.

Beam width

Most radar equipments develop their beam by illuminating a large specially shaped reflector from a small aerial in much the same way as the filament illuminates the reflector of a car headlamp. The analogy is very close, as light is also an electromagnetic radiation but of about one millionth of the wavelength of radio or radar. Unfortunately the basic laws of physics determine the width of the beam emitted by such an arrangement, this being inversely proportional to the ratio of the reflector size to the wave length in use. For equivalent sharpness to the beam associated with a normal car 10 cm diameter spot lamp, the radar aerial would have to be a 10 km in diameter!

The beam of a radar aerial does not in consequence have such a sharp cut-off at the edges and even narrow beam equipment may have measurable radiation 45° or more from the axis of the beam. The width is therefore defined as the angle between those directions in which the power transmitted is half that radiated along the axis of the beam. Using this definition an approximation of aerial beamwidth may be determined from the formula:

$$BW \approx \frac{65\lambda}{D}$$

where

BW = beamwidth in degrees

D = aperture of aerial } in the same units

λ = wavelength in use } in the same units

It also follows that as some energy is radiated outside the defined beamwidth of the aerial, in consequence some returns would be received. However, as the power radiated on-beam is considerably higher than that at other angles, the received returns would, in general, be from nearby objects at ranges well within the effect of the swept gain control.

Whilst considering beamwidth calculation it is also useful to remember than $\sin 1° \simeq \frac{1}{57}$, or in practical terms, two targets both at a range of fifty-seven miles spaced by one mile will subtend an angle of 1° at the aerial head.

This formula is correct to within one minute of arc. An alternative but less accurate form is that a spacing of 100 ft subtends an angle of 1° at 1 nm.

All discussion so far has assumed the consideration of a surveillance radar aerial for which the requirement is a narrow beam in azimuth. It is not, however, desirable–except in certain specialised circumstances–to limit the beamwidth so drastically in the vertical plane as high flying aircraft may well subtend high angles of elevation. It would therefore appear that the appropriate vertical response should give complete coverage from horizontal to vertical. In practice, this could cause difficulties at the limits of cover. Extremely low radiation, below 1°, would cause excessive ground returns, from hills, large buildings, etc., while excessive high angle (above 45°) radiation would be wasteful in power as any target aircraft at that angle of elevation is unlikely to be more than 15 nm distant. The ideal is to approach as closely as possible to a $\text{cosec}^2\theta$ pattern which will give reasonable vertical cover while maintaining maximum range.

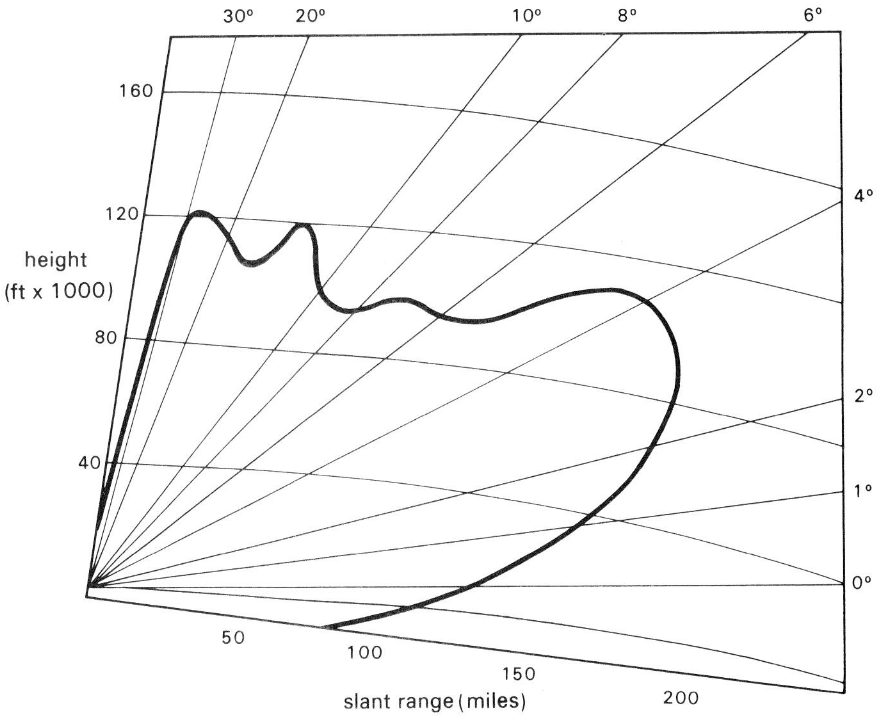

Fig. 32 Vertical polar diagram of a typical surveillance radar

Aerial positional information

The physical position of a stationary radar aerial may be determined by survey but when that aerial is rotating at speeds of perhaps 15 rev/min a more elegant method must be sought. Of equal importance is the requirement that this information must be made available at the radar display, which, in modern air traffic systems may be several hundred miles distant from the radar head.

Where the aerial head and the operational display are in close proximity such as within an airfield perimeter, use is often made of simple a.c. or d.c. servo systems. These require several pairs of interconnecting wires between the 'master' and 'slave' units and although this is quite practical over relatively short distances, the economics of renting several pairs of wires for turning information only, over the distances involved between remote radar stations and air traffic control centres, together with problems arising from noise developed on these lines, dictate that an alternative system be sought.

The development of digital techniques has led to the advancement of what is known as digital turning information. In this system an encoder is attached to the aerial turntable which generates 4096 pulses (known as Azimuth Count Pulses or ACPs) for each aerial rotation. These ACPs are fed down one pair of lines to the air traffic control centre where they are used to synchronise the displays.

Also attached to the turntable of each radar equipment is a striker which energises a switch at one specific point during the aerial rotation. This is normally located such that the switch is made momentarily while the aerial is pointing north and is consequently called the 'north marker'. The circuit energised by this switch is connected to the radar display where it causes an indication of instantaneous aerial position suitable for an alignment check.

Frequency diversity

It is normal practice with navigational aids for two separate sets of equipment to be installed at each site, one being held in reserve to take over immediate service should the operational transmitter fail.

In radar installations two equipments are again installed but in this case it is normal for both equipments to be operated simultaneously.

The two transmitters operate on different frequencies, these being displaced by at least 3%. The more the frequency difference, the greater the advantage until this difference exceeds the system's ability to use units or components of a common frequency bandwidth. To reduce the possibility of arcing within the waveguide feeder, the pulses are displaced in time by about 2–5 microseconds.

Pulse realignment is achieved by inserting a delay line corresponding in time to the transmit pulse displacement in the receiver path of the equipment emitting the initial pulse.

This method of operation offers three main advantages:

1. The pulse length and mean power of the radar equipment is effectively doubled without degrading the range definition, giving an improvement which increases with frequency difference and, practically, lies within the band of 15 to 30%.
2. As the reflectivity of an aircraft varies somewhat with frequency, the probability of a good return is higher than for single frequency operation, and the higher the frequency difference, the higher this probability.
3. Should a transmitter fail, although there will be a slight degradation in returns, no break in service will occur.

Pulse compression

From the foregoing, it may well be deduced that if a third pulse were added, a further improvement would be apparent. This is certainly so, but modern signal processing has enabled a far more elegant solution. This is the technique known as pulse compression. In this, a pulse of perhaps 40 microseconds length is radiated. This is also modulated so that the transmitted frequency varies continuously throughout the duration of the pulse, the overall shift being in the order of several tens of megahertz.

When a return is received, the receiver signal processing circuits insert a time delay which is dependent on the received frequency. Thus, if the pulse length is 40 microseconds, signals received on a frequency corresponding to the beginning of the pulse will be delayed by 39 microseconds. Signals on the frequency corresponding to the second microsecond of the pulse will be delayed 38 microseconds, and so on.

When all the signals are summated, they will give a strength of return corresponding to a 40 microsecond pulse, but a range definition appropriate to a one microsecond pulse, i.e. approximately one-tenth of a mile. Furthermore, the probability of return will be further enhanced by the diversity of frequencies employed.

Such improvements cannot, however, be gained without some penalty – in this case that the minimum range is limited by the duration of the pulse for, obviously, nothing can be received while the transmitter is operating.

In the example of the 40 microsecond pulse, the minimum range will be 40/12.36 or approximately 3.2 nautical miles.

Such a minimum range obviously cannot be tolerated, so a short pulse which may or may not be frequency modulated is radiated in turn with the long pulse which is processed by a separate receiver chain. The output of this is then superimposed on the main display.

Moving target indication

If a radar equipment is capable of detecting the relatively small signal return from an aircraft it follows that it will also be capable of detecting returns from hills, buildings or similar large objects which lie within its beam. Such returns can frequently cause embarrassment for their strength can easily mask the return from an aircraft flying above them. In consequence surveillance radar equipment incorporates circuitry designed to eliminate returns from stationary objects from the displays.

Two methods are in use for Moving Target Indication (MTI) the older using analogue, and the more recent, digital techniques. The basic principle of each system is that the echoes from one pulse are compared with those from its predecessor(s). In such circumstances echoes from fixed objects will remain constant in position whilst those from moving objects vary, thus by electronic subtraction, the fixed returns (called Permanent Echoes or PEs) may be eliminated. The difference between the two systems lies mainly in the method of storage of information in readiness for comparison, the older analogue method using a delay line and the more modern method a computer type memory circuit.

Analogue MTI

Although, almost certainly, the analogue MTI systems will eventually be replaced by their digital counterparts, such a replacement programme will be expensive and it may be many years before the system to be described falls into total disuse.

If the target aircraft is approaching or receding from the radar head, except for brief periods during which it is flying at a tangent, the echoes must show a frequency shift proportional to the relative velocity of the aircraft and the radar head. This is due to the Doppler effect and as echoes from fixed objects will show no such Doppler shift, this effect may be used to differentiate fixed from moving targets.

This moving target frequency shift can be readily detected if the receiver local oscillator is phase-locked to the transmitter. The output due to the moving target will be a series of echo pulses whose amplitude varies, within an envelope of the Doppler frequency, as the returned echoes become in and out of phase with the local oscillator.

It will be apparent that if, during the interval between two successive pulses, the aircraft approaches exactly one or more half wavelengths, the echoes will be indistinguishable from those due to stationary objects. Likewise, an aircraft flying at a tangent to the radar head will return successive echoes in the same phase and similarly be indistinguishable from fixed targets and will not show up on a moving target indicator. The approach speeds at which cancellation will occur are known as 'blind speeds' and loss of signal due to flying a tangential path is called 'tangential fading'. Further discussion on 'blind speeds' and methods of reduction of its effects will be found under 'Staggered PRF'.

To detect the phase shift between the echoes from target from successive

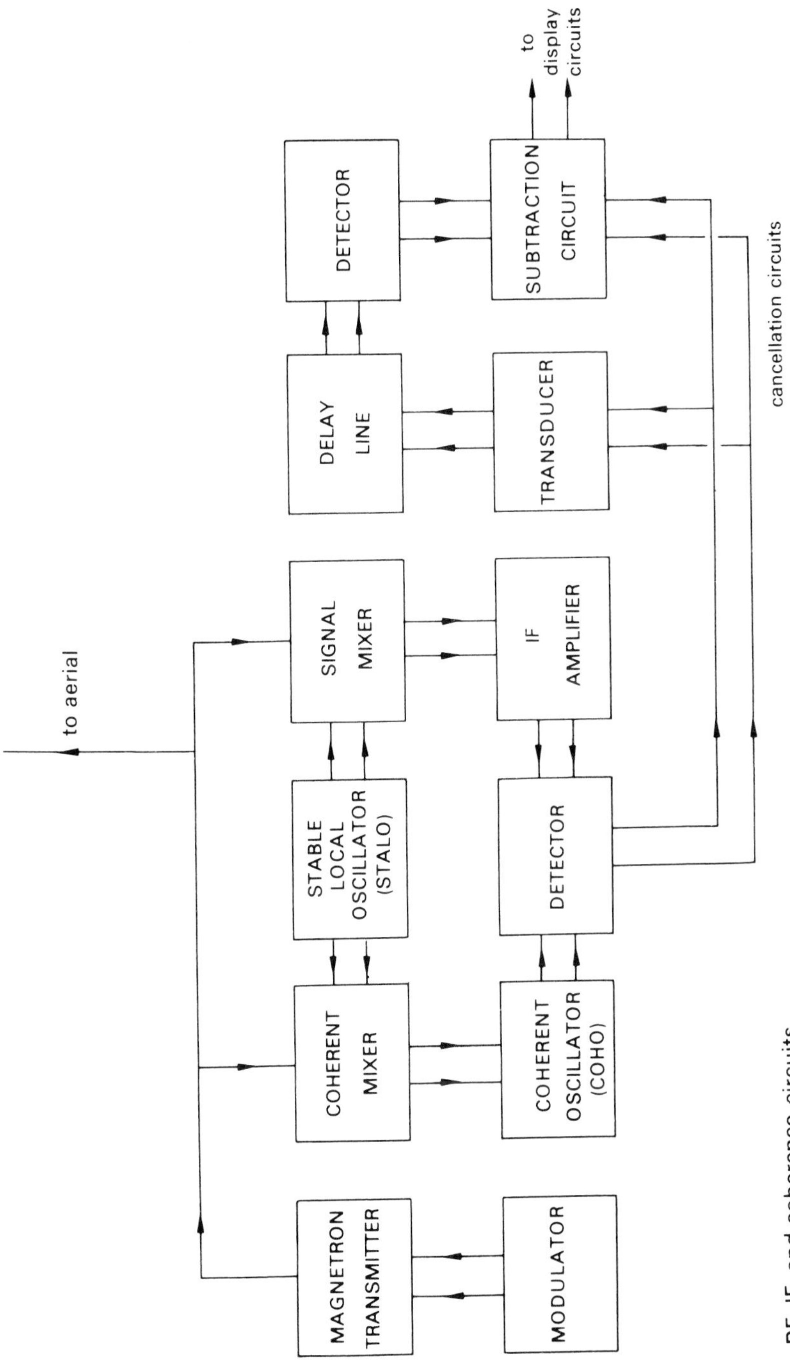

RF, IF, and coherence circuits

Fig. 33 Analogue moving target indicator radar–schematic

pulses it is necessary that the pulses from the transmitter all commence at the same phase. With transmitters using Klystron power amplifiers driven by high stability driver units this is possible, but with transmitters employing a magnetron (a high powered oscillator valve) this is not possible and an alternative approach has to be made. This method accepts that the transmitter will start with random phase at each pulse and arranges that a reference oscillator, which may alternatively be used in the phase comparison process, is phase-locked to the transmitter at each pulse.

With either system the receiver output will comprise permanent echo returns which will remain constant in phase and amplitude and target returns whose phase and amplitude will vary from scan to scan. If the return from one scan is subtracted from the return from the succeeding scan, the permanent echoes will disappear and only the moving targets remain.

To achieve this scan by scan substraction the output of the receiver is split into two paths, one direct and the other of such a length that the time of transit of the signal requires one recurrence period. The two signals are then combined in a balanced detector where the subtraction is performed.

As the velocity of radio waves is 300×10^6 m/s and the time between successive pulses is possibly in the order of 2000 microseconds, the length of the delayed path, in free space would be:

$$\frac{2000}{10^6} \times 300 \times 10^6 = 600\,000 \text{ m}$$

Such a path length is obviously impractical to incorporate within the equipment. Use is therefore made of an ultra-sonic transmission line, known as the 'delay line'. These are made of either mercury (in early equipments) or quartz. The receiver output signal is applied to a transducer which converts it from an electromagnetic to an ultrasonic signal, this transverses the material of the line at a relatively slow speed and after the required time delay, which is determined by the physical length of the line, is restored to an electromagnetic signal by a further transducer. In early equipment a mercury delay line was used which took the form of a long tube filled with mercury with transducers fitted at either end. The delay characteristic of this type of delay line varies with temperature and the degree of impurity present (i.e. dust). It has now been almost completely superseded by the quartz delay line.

Although similar in principle to the mercury line, the quartz delay line differs considerably in design, basically because quartz is a solid, unlike mercury, which is a liquid.

This enables the quartz to be cut in the form of a polygon so that from the transducer the ultra-sonic signal traverses the quartz, impinges on an opposite face, which by its incident angle reflects the signal to a further face and so on until it finally reaches the output transducer. The time of the delay is dependent on the length of path in consequence of having traversed the quartz many times due to the reflective process.

The use of the delay line necessitates strict control on the PRF of the equipment, any variation between the delay of the line and the period between consecutive pulses causing a severe degradation of MTI performance.

It is normal practice in equipments using MTI, for the modulator (which controls the 'firing' of the transmitter) to be controlled by a trigger pulse from an oscillator. This is generated a brief time before the transmitter pulse is due and as far as the delay line is concerned occurs at a time at which there is no other input. The trigger pulse can therefore be led through two paths–one to energise the modulator and a second to the delay line. After passing through the delay line that pulse should be coincident with the next trigger pulse. Thus the timing of the two pulses can be compared with the error signal being used to correct the periodicy of the trigger oscillator.

We have already seen that in order to recognise pulse-to-pulse phase changes it is necessary that the returning signals be compared with an oscillator whose phase is locked to each transmitted pulse. Such a phase-locked oscillator is said to be 'coherent'. The coherent oscillator may operate at either transmitter or IF frequency, the latter being the more usual.

In this technique a small fraction of the transmitter output is mixed with the output from a very high stability local oscillator, known as the 'stalo' (*Sta*ble *L*ocal *O*scillator) to produce a pulse which is used to reset the phase of a stable oscillator at the intermediate frequency. This is referred to as the coherent oscillator or 'coho' and its input pulse the 'coho locking pulse'. The incoming echoes are also mixed with the stalo and after amplification the resultant IF signals pass to a Phase Sensitive Detector (PSD) in which the coho acts as the reference oscillator. It is likely that the phase of the transmitter varies from pulse to pulse, but the phase of the returns received from fixed targets will vary in like manner. As both the transmitted and received pulses are mixed with the same oscillation, the phase of the coho and that of the IF signals in the receiver vary in a similar way provided that the stalo is sufficiently stable over the period between the emission of the pulse and the reception of the echo. Due to this identical pulse-to-pulse phase variation in the coho and echo signals, the amplitude of the video output due to fixed targets will be constant but the additional phase variation due to moving targets will not be shared by the coho and a varying output will result.

If the transmitter uses a low power oscillator driving a Klystron amplifier it is possible to lock the transmitter phase to the coho rather than vice versa. This can be achieved by mixing the outputs of the coho and stalo to produce a signal which after amplification will drive the Klystron high power output stage. Other combinations have been evolved such as RF locking with IF phase comparison or IF locking with RF phase comparison but no matter which system is used the purpose is to generate returns from fixed objects which are of constant strength while echoes from moving targets are of varying amplitude.

Clutter switching

In the simplest arrangement, video from an additional linear detector operating in parallel with the phase sensitive detectors is compared against an amplitude threshold. Any crossings which exist longer than a preset time generate a gating waveform. This is used to switch in MTI video for regions of clutter lying outside the preset MTI range gate.

Although the system is effective against unbroken clutter it is not effective against broken clutter and in order to overcome this problem a more complex form of detection has to be employed. In this, the percentage area of each sector wherein there are returns greater than threshold is determined by monitoring the linear video. This data, modified if necessary by data concerning the surrounding areas, determines whether or not MTI data should be used in the area of interest. The clutter data is put into a store until the next radar revolution, when it is used to select the correct video.

Pulse recurrence frequency discrimination (PRFD)

This process removes interference due to other non-synchronised radars. The video signal is compared against a threshold level, any crossings being passed to a small store. Both current incoming and delayed crossings are compared and if the former is present without the latter, then the current video is blanked to zero.

Digital moving target indication

The signals to be processed in the digital MTI system are obtained from the radar transmitter/receiver as an intermediate frequency input and to ensure optimum performance the signal/noise ratio is matched to system requirements by use of a compression amplifier housed in either the transmitter/receiver or the processor.

The compressed intermediate frequency signal is taken to two phase-sensitive detectors in parallel. These are similar except that the phase of the coho reference signal to one is shifted by 90°, this method substantially reducing 'blind phases', that is when the absolute phase of the incoming signal would result in zero output if only a single phase-sensitive detector (PSD) were used.

The analogue signal from each PSD is then quantised to digital form by an analogue-to-digital converter operating at a sampling interval dependent on the transmitter pulse length. This is then fed to two computer-type stores in series, each capable of holding the information received from one pulse of the transmitter. At the end of each full range count therefore, the stores hold the data from that and the preceding count. At the start of the next count these two sets of data are shifted out and a new set of data shifted in. The three sets of data are presented to a comparator, or summation circuit, in which twice the once-delayed data are subtracted from the sum of the undelayed and twice-delayed data.

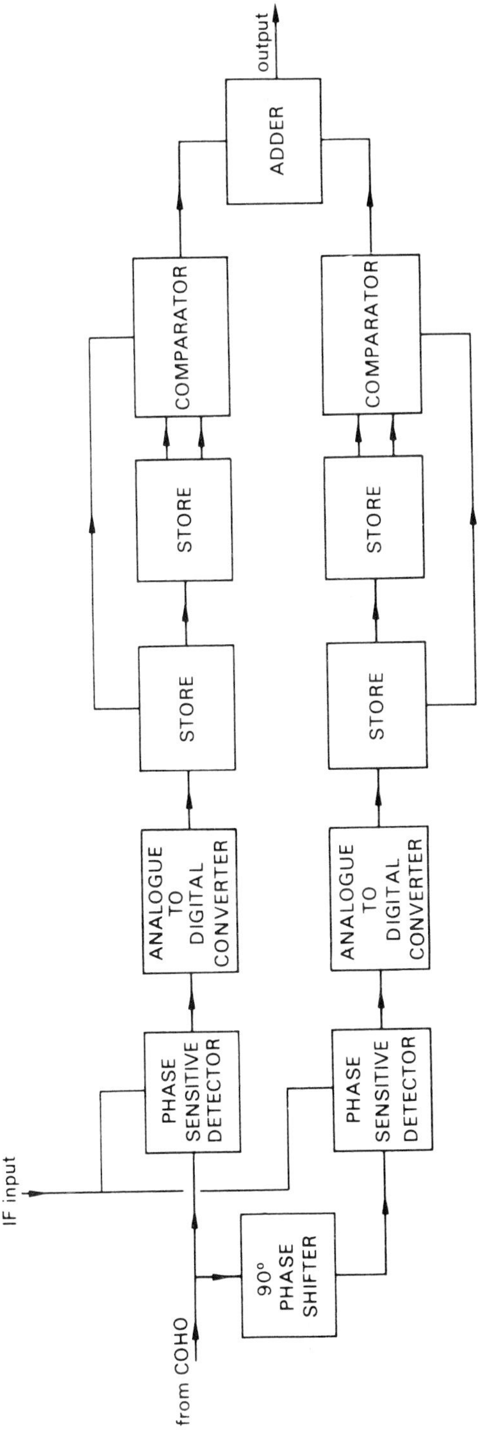

Fig. 34 Digital MTI-block schematic

The outputs of the two channels are then passed to a simple adding circuit to produce a single video signal. When the system is using frequency diversity, the output from the other signal processor may also be combined at this point. This ensures the best possible signal for further processing.

The use of digital techniques greatly facilitates processing such clutter switching, Pulse Recurrence Frequency Discrimination (PRFD) and de-staggering.

An MTI system such as that described above has a frequency response which rejects signals of zero Doppler frequency (i.e. fixed targets). Its bandwidth is narrow and thus preserves signals from targets with low radial velocity. It is also possible to create Doppler filters at frequencies other than zero and use them for 'acceptance' rather than 'rejection'. A bank of these (typically eight) can be created which not only allows fixed targets to be filtered out, but also allows the radial velocity of moving targets to be measured. This is helpful in subsequent tracking computer operations. Such an arrangement is usually referred to as Moving Target Detection (MTD).

Staggered PRF and fading

Reference has already been made to the effect that if the target aircraft moves on an exact number of half-wavelengths, at the radar frequency, to or from the radar head in the time interval between two consecutive pulses, the echo returned to the radar head from each pulse will be in the same phase and in consequence appear as a permanent echo to the MTI circuits with inevitable cancellation.

The period between two consecutive pulses is $\frac{1}{\text{PRF}}$ seconds and if the aircraft is approaching the radar head at a radial velocity of V m/s, the distance traversed by the aircraft in $\frac{1}{\text{PRF}}$ seconds will be $\frac{V}{\text{PRF}}$ and to calculate the lowest 'blind' speed this must equal the radar half-wavelength $\left(\frac{\lambda}{2}\right)$. Therefore

$$\frac{\lambda}{2} = \frac{V}{\text{PRF}}$$

Considering a radar with PRF = 500, $\lambda = 10$ cm, then

$$\begin{aligned} V &= \frac{\lambda}{2} \times \text{PRF} \\ &= \frac{0.1 \times 500}{2} \\ &= 25 \text{ m/s} \\ V &= 90 \text{ km/h} \quad \text{or approximately 48.5 knots.} \end{aligned}$$

Successive blind speeds will also occur at all multiples of this velocity, i.e. 97 knots, 145.5 knots, 194 knots, 242.5 knots, 291 knots, etc. Such 'blind' velocities could obviously cause considerable embarrassment to the controllers operating the radar. In the example quoted there are no less than ten different speeds relative to the radar at which a target aircraft will be invisible within the normal sub-sonic flight envelope. Furthermore, as the blind speeds are relative to the radar head it is possible for a fast aircraft flying at around six hundred knots in a straight line across the area covered to fade no less than ten times approaching, suffer a tangential fade whilst passing and a further ten times receding from the radar–an obviously untenable situation.

We have already demonstrated that the blind speeds are a function of PRF, thus if the area previously referred to was covered additionally by a radar with a different PRF, the aircraft would be visible on at least one radar at all times unless the approach speed coincided with a velocity which was a blind speed for both radars. Although possible, such a solution would obviously be undesirable on economic if no other grounds and means must be devised to arrange that one radar can be operated at a number of PRFs simultaneously.

If a random 'jitter' were permitted on the recurrence frequency in effect the requirement for varying PRF would be met but this would render the MTI inoperative. Recourse is therefore made to delaying the transmission of alternate or every third pulse by a fixed time of typically fifty microseconds. Realignment of the video signals for MTI purposes can be achieved by switching the signals of non-delayed pulses through a delay line of time equivalent to that pulse delay.

If alternate pulses are delayed, then for a nominal PRF of 500 with a delay of fifty microseconds:

Time between first and second pulses = 2050 microseconds i.e. PRF = 487

Time between second and third pulses = 1950 microseconds i.e. PRF = 513

The transmitter is therefore operating with two PRFs of 487 and 513 pulses per second. Should only every third pulse be delayed, the spacings between pulses will be 2000, 2050, 1950, 2000, . . . microseconds, equating to PRFs of 500, 487, and 513. On a 10 cm radar these PRFs would be susceptible to blind speeds of 48.5, 47.3 and 49.8 knots respectively and the lowest blind speed of the combination would be the lowest common multiple of the three. It is also possible, and it is used in some equipment, to use a 'random' stagger provided that the system knows in advance what the 'random' pattern is to be.

Adaptive signal processing

In addition to the MTI and MTD techniques so far discussed, a third system has recently been introduced, which enables the principle of Doppler filtering to be used in a cheaper form. This is known as Adaptive Signal Processing (ASP).

In this technique the incoming signal is split into three simultaneously operated parallel processing channels:

1. A non-coherent detector
2. A moving target indicator with zero Doppler filter
3. A moving clutter rejector (auto-Doppler) with an automatically variable Doppler frequency filter.

After Temporal Threshold Integration, the output of the three channels is applied to an OR gate so that any target 'in the clear' in either the fixed or moving clutter is selected and thus made available for display.

The auto-Doppler channel senses the Doppler frequency of any moving clutter and sets a rejection notch which automatically follows the clutter area.

Height finding radar

Most height finding radar equipments operate in exactly the same way as their surveillance counterparts except that the beam transmitted from the aerial is narrow in the vertical plane, broader in azimuth and sweeps through a vertical angle from horizontal to 45° or 60°. The height of the target aircraft is determined by comparing range with angle of elevation. However, when it is realised that two aircraft, flying at standard airways vertical spacing (1000 ft) at a distance of sixty miles from the radar will only differ in elevation by ten minutes of arc, some of the difficulties associated with the system can be realised. In civilian ATC practice it is generally considered preferable to rely on the SSR mode C (height) readout.

The only exception to this in recent years is in precision approach radar which incorporates an elevation element. However, as the operational range is not excessive and the wavelength in use 3 cm, adequate accuracy is obtainable.

An alternative method of height finding is to switch the aerial array so that two different vertical polar patterns are produced. These differing patterns cause two differing strength returns from each target aircraft and a comparison of these signal strengths allied to a knowledge of the VPDs can determine the aircraft altitude. This system is known as split-lobe height finding.

Modern military height-finding radars utilise multiple stacks of beams (typically eight) in the elevation plane. Signals from a given target in contiguous beams will be of differing amplitude. A pre-knowledge of the beam shapes allows the amplitude differences to be converted to an elevation angle which, together with the target range, permits height to be calculated. These, however, are not as reliable in absolute terms as SSR Mode 'C' values.

Radar plot extraction

Air traffic control techniques are becoming increasingly dependent on the use of radar. In consequence, the radar system feeding this information to the control centre must in itself be inherently reliable and failsafe. This can be achieved by multiple coverage of the required airspace using a number of different radar stations. When this information is received at the air traffic control centre it is combined to form a single system.

In order to process the vast quantity of data received from such radar systems, it is necessary to equip the associated air traffic control centre with a major computer system, which, in addition to processing and combining the incoming information, must present the air traffic situation on specialised displays showing both geographical and alpha-numeric data.

To permit the radar equipments, whose output is analogue signals, to operate in conjunction with computers whose input requirement is digital signals, a sub-system must be included whose purpose is not only to 'digitise' the analogue radar signals but also to subject these signals to digital pre-processing.

As a result of linking the radar and computer, certain tasks previously performed by the radar controller may now be performed automatically. These include the detection of valid targets, determination of target position and tracking of target movement. This the radar controller achieved on the basis of the distribution of the radar echoes displayed, discriminating between aircraft targets and clutter caused by ground and weather returns by experience. Memory and grease pencil marks on the face of the display tube provided the target tracking information.

In an automated system, target tracking from fed-in target position data can best be performed by a general purpose computer. It can, therefore, be accomplished additionally by the computer provided for general information processing.

For target detection and determination of target coordinates, it is however, preferable to utilise a system specifically designed for the purpose which assumes the function of connecting link between radar and computer. This system is called the radar plot extractor.

Primary radar plot extraction

The purpose of the primary plot extractor, which is located at the radar station, is to accept the output from a radar MTI system, check the validity of each return and for valid plots convert the rho/θ format of positional information to cartesian coordinates in a digital form suitable for feeding via a modem and a narrow band link (e.g. a normal telephone line) to the associated air traffic control centre.

The incoming signal is initially fed to strike extraction circuits where it is examined for signal level and pulse length. If the incoming signal is being received from an analogue MTI system, it is converted to digital form at this point. Signals meeting the required criteria are then passed on to the plot forming circuits.

The function of plot forming is to correlate the strike received in successive pulse recurrence frequency (PRF) periods through an interval corresponding to the radar aerial beam width, thereby building up complete target plots.

Noise pulses are largely eliminated at this stage by their failure to correlate in range. Clutter returns are filtered by checking plots against predetermined azimuth criteria.

The plot data are stored in range order for checking against preset acceptance criteria and strike data. Each time a strike is registered, the partially-formed plots in the store are compared with predetermined criteria and optimised for a particular signal structure. The predetermined criteria are:

(a) Leading edge–minimum strike pattern to initiate a plot.
(b) Trailing edge–miss pattern to terminate a plot.
(c) Run length–the maximum allowable strike pattern.
(d) Range variant–maximum and minimum limits for range correlation.

The range of the incoming strike is compared to the current partial plot range. If it is less than the variant a new partial plot is initiated; if it is within the variant a hit is declared, the plot data are updated and the leading edge and run length criteria are applied; if it is greater than the variant a miss is declared and the trailing edge criterion is applied. If the criteria of leading edge and trailing edge but not run length are met, then a valid plot is declared, otherwise it is declared void and the store cleared.

Plot bearing calculation

The angular position of the aerial is continuously obtained from the digital shaft encoder. If these data are combined with the time period defined by the PRF of the radar, a measure of the aerial angular velocity can be found. This, combined with the absolute angular position and the width data for a valid plot, enables the bearing of the plot centre to be calculated. By performing the velocity calculation over a large number of traces, bearing errors due to aerial rotation speed fluctuation can be reduced to less than the angle turned between traces and PRF stagger can also be accommodated.

The output of this unit is a digital form suitable for connection to a computer via a suitable interface or to a link buffer unit for combination with secondary radar plot, aerial bearing and status data for driving the modems associated with the telephone lines connecting the radar station to the air traffic control centre.

The ability to use standard telephone lines for this purpose also gives considerable financial advantage over the use of low loss coaxial cables or microwave links which have to be used for the transmission of analogue signals over long distance.

5.2 Secondary surveillance radar (SSR)

The development of secondary radar stems from the wartime radar controller's requirement to be able to differentiate between the radar returns from enemy or friendly aircraft.

In its most elementary form, IFF (Identification, Friend or Foe), as it was then called, consisted of a transmitter operating on about 200 MHz feeding an aerial mounted on top of the primary radar aerial, interrogating a simple transponder fitted to the aircraft.

The IFF returns were superimposed on the primary radar display, the effect being to cause a substantial thickening of the displayed echo, thus facilitating identification.

Subsequent development showed that where the aircraft is fitted with a suitable transponder, secondary radar has three major advantages over primary:

(a) The transmitter need be of only relatively low power compared with primary radar equipment.
(b) The returns, not being dependent on reflection, but consequent of a transmission from the aircraft, are of superior signal strength and improved reliability.
(c) The returns from the aircraft may be coded to pass information to the radar station.

These very advantages give rise to a series of problems.

As the strength of the signal return is high compared with primary radar, aerials tend to be of minimum size compared with their primary radar counterparts. This has enabled mounting on top of the associated primary radar aerial, thus eliminating problems of bearing synchronism.

A small aerial, however, has wide beamwidth in both the horizontal and vertical planes and the radiation pattern also has substantial sidelobe content.

Wide horizontal beamwidth can cause problems of poor target bearing definition, and wide vertical beamwidth results in heavy ground returns, whilst both can cause problems from false replies due to reflections from buildings, mountains, etc. Similarly, false returns could also be received due to interrogations by sidelobes of the main aerial pattern.

These problems have been minimised by a combination of aerial design and signal processing techniques.

Side lobe suppression

The earliest of these were incorporated in the basic SSR concept and both addressed the sidelobe problem and instructed the aircraft transponder of the information required.

This is achieved in the method of interrogation which comprises three pulses, designated P1, P2, and P3, of which P1 and P3 are radiated on the main beam of aerial array. P2, known as the SLS (Side Lobe Suppression) control transmission pulse, is radiated omni-directionally at such an amplitude that it is equal to or greater than the signal strength of the greatest side lobe transmission of the aerial radiating P1. Furthermore, the level of P2 is arranged to be a level lower than 9 dB below the radiated amplitude of P1 within the desired arc of interrogation. A typical method of producing the required horizontal polar diagrams is by use of a 'sum and difference' aerial in which the interrogate pattern has the normal main lobe/side lobe structure but the control (P2) pattern has a substantially omni-directional structure but with a null in the direction of the interrogate main lobe. In this aerial, the radiating elements are divided into two halves. If both halves are fed with equal power in phase, the signal from the two halves of the aerial combine to develop the narrow beam necessary for the P1 and P3 pulses. This is known as the 'sum' pattern. If the phase of one feed is reversed, the two individual patterns interfere with each other producing a difference or SLS control (P2) pattern which is virtually omni-directional.

The design of the aircraft transponder is such that no reply will be made unless P1 exceeds P2 by at least 9 dB. Where P2 is equal to or stronger than P1 the transponder receiver is muted for a period of thirty-five microseconds. This effectively eliminates any possibility of interrogation by side lobe radiation.

Monopulse

As stated previously, the beamwidth of the interrogating transmission is relatively wide and in order to determine the position of the target, the signal processing circuits of the ground equipment derive a mean position of the target as it is illuminated by successive 'strikes' from the interrogator.

Unfortunately, for a number of reasons, the strength of the return from succeeding interrogations is not necessarily constant, with the result that the indicated may not correspond with the actual target position.

When this occurs on successive aerial rotations, the track of the target aircraft may well be indicated as an irregular zig-zag, even though the aircraft is steering a constant course. This effect is known as track-wander.

This has been largely eliminated by the 'monopulse' technique which accurately measures the bearing for each individual returned pulse, as compared with the basic SSR system which determines bearing from the mean of a number of strikes. In this system, for reception, as for transmission, the aerial is divided into two halves, from which two separate outputs are derived: one corresponding to both parts in phase (sum) and the other, antiphase (difference). These are then fed to separate receiver chains, the phase and

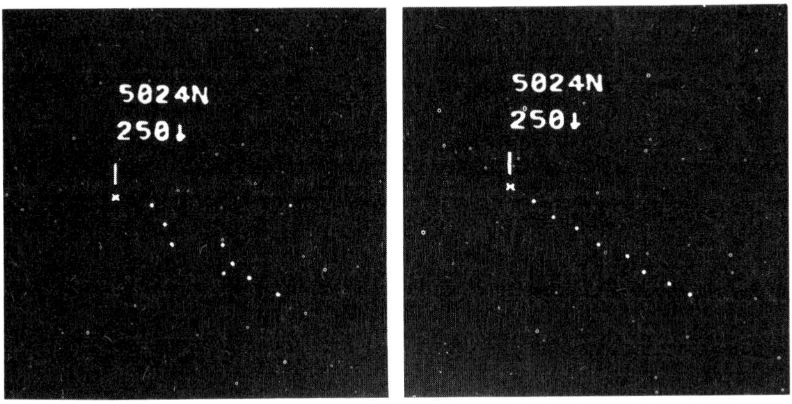

Plate 26 Comparison of track-wander with (left) conventional and (right) monopulse SSR systems. *(Photo: Cossor)*

amplitude of the output being compared to derive the off-boresite bearing of the target.

The improvement in accuracy by using this technique may be realised if an example is considered. If an interrogated aircraft gives 25 replies, each of eight pulses in one beamwidth, the traditional technique will derive the bearing from two of these (the first and the last). The monopulse system, in comparison, will derive a mean bearing from $8 \times 25 = 200$ determinations – an obvious improvement.

LVA (Large Vertical Aperture) antennas

The standard SSR interrogator aerial normally has a horizontal aperture of about 4 metres and a vertical aperture of about 0.5 metres. Due to their shape, these are commonly called 'hog-trough' antennas.

Plate 27 A Cossor LVA (Large Vertical Aperture) SSR aerial. This is also designed for use with Monopulse systems. *(Photo: Cossor)*

Because of this small (about 1½ wavelengths) vertical aperture, the beam radiated in the vertical plane is very broad. As a result, the ground in the nearfield is heavily irradiated and the consequent reflections interfere with the main beam structure causing heavy lobing in the vertical polar diagram. This results in the track history of a target passing through successive lobes and nulls being discontinuous. Furthermore, a considerable proportion of the radiated signal is at high angles where no targets can possibly exist.

These problems have largely been overcome by the introduction of the LVA (Large Vertical Aperture) antenna. In this, the vertical aperture has been increased to about 6 wavelengths (1.6 metres) which results in a considerable reduction in vertical beamwidth. Furthermore, by varying the phase and amplitude of the signal fed to the radiating elements, tailoring of the vertical polar diagram (VPD) is possible, reducing output below horizontal and at high angles, with consequent enhanced radiation at elevations at which it is most effective. The reduction of radiation at elevations below horizontal also minimises lobing and consequent track history dicontinuities.

The interrogation

In the pulse pattern radiated by the transmitter, the interval between P1 and P2 is two microseconds but the position of P3 is variable, this being the method by which the type of information required from the transponder is indicated. Four 'modes', or types, of reply are at present allocated for civil use, designated A,B,C and D. The pulse spacing corresponding to these modes and their current applications are:

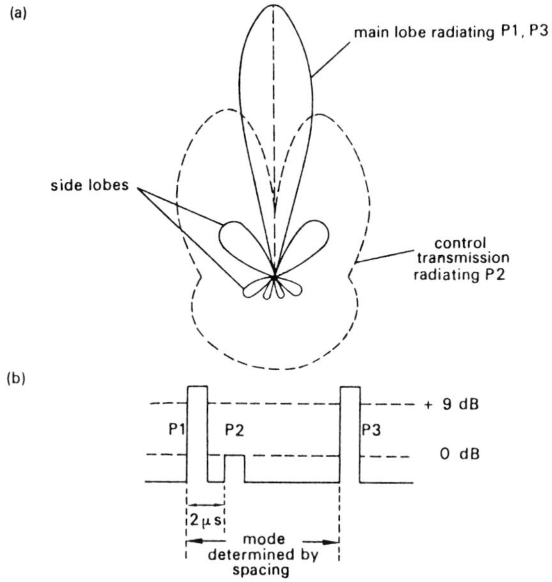

Fig. 35 SSR radiated (a) horizontal polar diagrams (b) interrogate pulse pattern

Mode	P1 to P3 spacing (μ s)	Application
A	8 ±0.2	ATC
B	17 ±0.2	ATC
C	21 ±0.2	Pressure altitude
D	25 ±0.2	Unassigned

On receiving a valid interrogation the aircraft transponder will radiate two framing pulses spaced 20.3 microseconds between which, at 1.45 microsecond increments after the first framing pulse, are the information pulses.

Between the framing pulses up to thirteen information pulses may exist. These are divided into four groups of three plus an additional position, designated X, which is at present unused and is specified only as a technical standard for possible future use.

Three pulses in each group corresponding to 1, 2 and 4, enable the numbers 0 to 7 to be coded in binary notation, thus with four groups the numbers 0000 to 7777 may be coded in BCO (Binary Coded Octal). This corresponds to 0–4096 in the more common decimal notation. The meanings of these codes is entirely dependent on the SSR mode in operation. The mode A reply represents the aircraft control identification, having been allocated by Air Traffic Control and entered into the aircraft equipment by the aircrew. The mode C (height) reply is completely automatic, operating from a pressure capsule within the equipment.

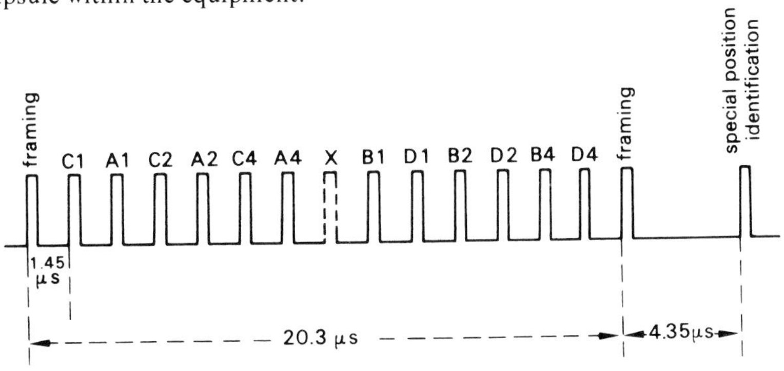

Fig. 36 SSR reply pulse patterns

In addition to the information and framing pulses previously described, the standard SSR format also allows for a Special Position Identification (SPI) pulse. This may be transmitted in combination, with any other information and is 'keyed' by the aircrew at the request of the air traffic control authority. The pulse is situated at an interval of 4.35 microseconds following the second framing pulse.

All secondary surveillance radar systems operate on the same frequencies, interrogation being at 1030 MHz and transponder replies being at 1090 MHz. This commonality gives rise to further difficulties. Although ground based interrogators may be geographically separated by over 200 nm, some aircraft may be within range of two or more stations and cause non-synchronous interference to one station by replying synchronously to another. This interference is termed 'fruit'. Although the ground receiver incorporates circuitry for separating synchronous from non-synchronous returns, 'defruiters', interference difficulties still arise. The probability of obtaining a reply free of 'fruit' decreases at a greater rate than the increase of transponders in use.

A further problem arising is that of 'synchronous garbling' which results from the overlapping replies of two or more aircraft in close proximity within the same aerial beam width. Typically, this occurs with overtaking, aircraft on airways and in holding or stacking patterns.

It is reasonable to surmise that although these problems are not insurmountable at present, air traffic level forecasts for the future envisage that air traffic increase in certain areas will be such that the present SSR system will, in time, become saturated by a proliferation of garbling and 'fruit'. It was with this situation in mind that the UK and the USA each commenced a development programme of improvements to the SSR system. In 1974 a series of discussions between the United States Federal Aviation Administration and the United Kingdom Civil Aviation Authority resulted in a Memorandum of Understanding (MOU) to bring compatibility to their respective systems. Furthermore, the MOU allowed for future developments of these systems to take place on a co-operative basis.

Mode S

The deficiencies of the present SSR will be largely overcome by the use of Mode S.

The requirements for the successor to the present SSR system may be summarised broadly as follows:

(a) elimination of synchronous garbling.
(b) increased capacity in terms of traffic numbers.
(c) improved surveillance accuracy to meet future demands of automated air traffic control systems.
(d) compatibility with existing ICAO SSR as described in Annex 10.
(e) minimal costs to aircraft operators through evolutionary developments.

The Mode S ground equipment operates on the same frequencies as SSR and comprises an interrogator and a receiver. Monopulse techniques are invariably used and this results in the off-boresight azimuth (OBA) of the interrogated aircraft being within five minutes of arc. Furthermore, range measurement is predicted to have an accuracy of approx 100 ft due to improved transponder design.

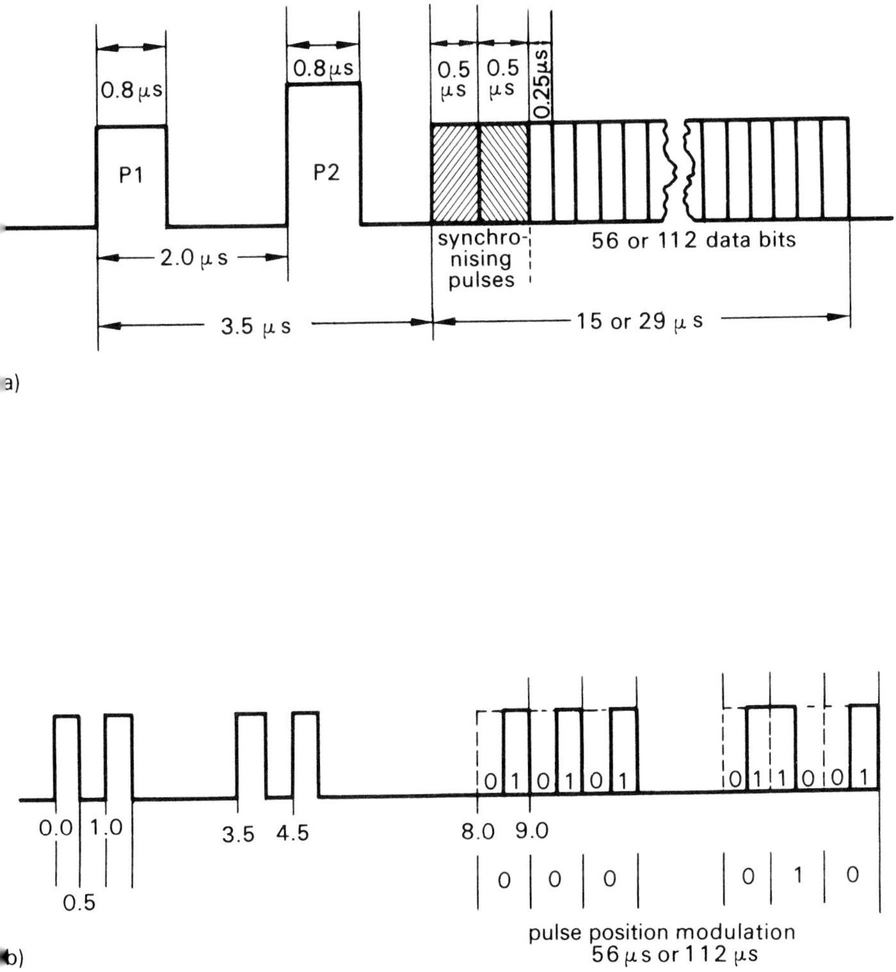

Fig. 37 Mode S (a) interrogation (b) reply

In addition to Mode S functions, the ground station will also radiate standard SSR modes and will therefore be capable of operating in conjunction with aircraft carrying standard SSR equipment. In the same way the Mode S transponders will be compatible with SSR ground stations.

The Mode S transponder is based generally on the type of unit in commercial airline operation. However, the characteristic which in an SSR transmission mutes the receiver is the one that identifies a Mode S transmission to a suitably equipped aircraft. The necessary relationship between the received levels of the P1 and P2 pulses for an SSR transponder to reply has already been described. In contrast Mode S interrogations are prefaced by a P1, P2 signal in which P2 is of greater amplitude than P1. This is followed by a one microsecond phase reversal synchronising signal and a message of up to 112 data bits (bit= Binary digIT). As this message includes the aircraft address, only the relevant aircraft will reply. The allocation of twenty-four bits for aircraft address permits over sixteen million discrete addresses–sufficient for individual registration of all aircraft throughout the world! Non Mode S equipped aircraft will recognise the P1/P2 relationship as a standard SLS (Side Lobe Suppression) transmission and consequently be muted.

The aircraft reply begins with a synchronising preamble consisting of two pairs of 0.5 microsecond pulses. This is followed by a block of data bits over a period of 56 or 112 microseconds depending on whether 56 or 112 bits of data are to be transmitted. These data are transmitted using pulse position modulation at a rate of one megabit per second. In this system each data bit occupies one microsecond which is divided into two intervals. If a 0.5 microsecond bit occurs in the first interval, a binary '1' is indicated; if it is in the second interval then the information transmitted is binary 'O'.

In order to acquire further Mode S equipped aircraft a special roll call interrogation known as the SSR/Mode S ALL CALL is broadcast at intervals. Normal SSR transponders will reply using normal mode A or C codes, ignoring a P4 pulse which follows the leading edge of P3 by 1.5 microseconds. Mode S transponders, whilst capable of replying to normal SSR interrogation will recognise the P4 pulse as a roll call request and will reply with an 'all call' response consisting of the aircraft identity plus the capability of the onboard equipment.

Other interrogation messages are:

(a) Surveillance interrogation–sixteen control bits, sixteen altitude-echo* bits, twenty-four bits for address and parity used for position up-date.
(b) Comm-A interrogation–as above plus fifty-six bits of ground-to-air data interchange message. Longer messages may be accommodated by successive cycles of interrogation-response.
(c) Comm-C Interrogation. 112 bits for more efficient transmission of long ground-to-air data interchange messages. Eighty bits are used for data

**Altitude-echo is a means of informing a pilot of the flight level which his transponder is indicating*

purposes in each interrogation and up to sixteen Comm-C interrogations may be acknowledged with a single transponder reply.

On receiving these interrogations the transponder will reply with either:

(a) All Call reply which uses 56 bits which includes the aircraft discrete address to enable ground processors to include it in the aircraft file.

(b) Surveillance reply which uses fifty-six bits and is the normal reply when no air to ground interchange message is needed. It consists of thirteen bits for SSR identity and altitude, nineteen bits for control purposes and twenty-four bits for address and parity.

(c) Comm-B reply using 112 bits. As for surveillance reply plus fifty-six bit air to ground message.

(d) Comm-D reply uses 112 bits for transmission of long air-to-ground messages. Includes an eighty bit message and up to sixteen Comm-D replies can be sent as a single long response and acknowledged with a single interrogation. Comm-D cannot be used for position updating because it does not include aircraft Flight Level.

Plate 28 The Plessey ACR 430 X-band radar which has been designed specifically to provide approach guidance and general surveillance at the smaller civil and military airfields. The single antenna reflector is illuminated by twin horns fed from dual transmitters to provide two-beam coverage. *(Photo: Plessey)*

Plate 29 The Telefunken 23 cm radar aerial which, in conjunction with Signaal electronics, will form the backbone of the UK civil air traffic control network for the 1980s and 1990s. It is surmounted by a 'hog-trough' SSR aerial. *(Photo: Author)*

Plate 30 The Marconi Radar Systems S512 airfield surveillance radar. This equipment operates on 10 cm wavelength. *(Photo: Marconi Radar Systems Ltd)*

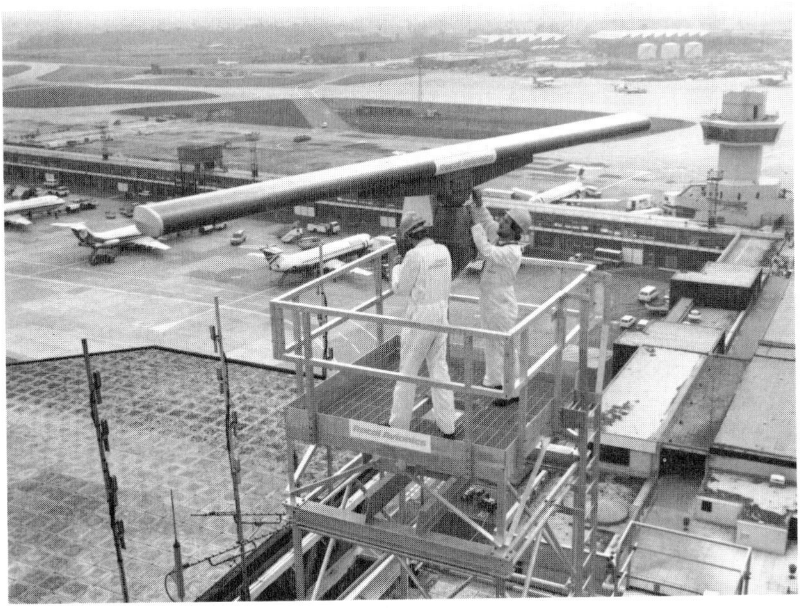

Plate 31 The aerial system for the Racal AR18x Airfield Surface Movement Indicator at Manchester Airport. *(Photo: Racal)*

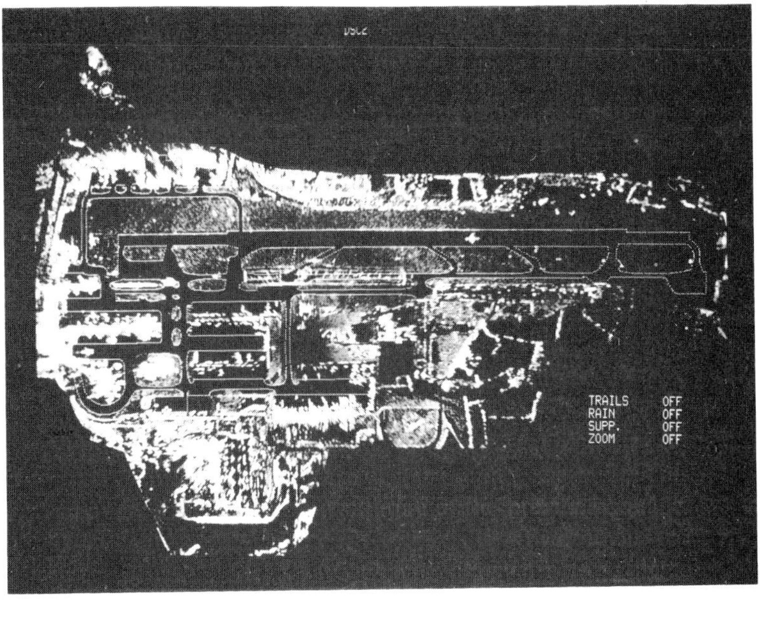

Plate 32 An off-screen photograph of the AR18x display at Gatwick Airport. Note the video map overlay, blanking of unwanted areas and the aircraft on the runway. *(Photo: Racal)*

Secondary radar plot extraction

Fruit and defruiters

The defruiter is used to inhibit unwanted non-synchronous replies present in the raw secondary radar returns. This process effectively minimises the display of spurious returns (fruit) and leaves a clear presentation of valid aircraft replies on the operator's display.

The basis of the defruiting process is a pulse to pulse comparison of relevant radar returns on consecutive scans utilising the same mode. If synchronous and therefore valid, a radar return appears at sensibly the same range for each range scan. After a few returns a pattern is established. The defruiter enables a selection to be made for returns to be compared from two, three or four consecutive range scans. Similarly, a selection can be made to define the number of these returns (two, three or four) that shall positively compare in order to indicate a valid SSR reply. This could be two positive comparisons out of two consecutive range scans, two or three out of three, two, three or four out of four. In practice, the criterion frequently employed is any two positive comparisons out of four range scans, i.e. if a return appears at a consistent range and is of the same code content for any two from a maximum of four consecutive range scans.

Within the defruiter the SSR mode is first detected to enable valid comparisons and the video is converted into digital form and fed sequentially into a computer-type store. This store operates on a read/write cycle and thus, on the fourth consecutive range scan, starting from any one instant of time, three stored words can be compared with the fourth word about to enter the store. This comparison is effected by correlation logic circuitry where the number of positive comparisons (agreements) are detected.

If the desired criterion has been met or exceeded as defined by the correlation logic, then the incoming video signal is allowed to pass to the decode unit.

The decode unit

The SSR inputs to the decode unit are fed directly to pulse threshold detectors, which establish the amplitude at which an input pulse is recognised as a valid input signal. Mode pulses identified here are transmitted through the decoding system as a binary code. The video, after threshold detection, is passed to the video conversion unit which standardises the amplitude of the video and processes it for entry into a twenty microsecond delay line. After leaving the delay line the pulses are reconstituted.

Framing pulse identification

The leading edges of both incident, F2, and delayed F1, pulses are used to initiate framing pulse identification. The framing pulse identifier uses the delayed F1 pulse, from the delay line, to generate a gate with an accurately controlled characteristic. This gating signal, F2 gate, is used to establish the presence of framing pulses, F1 and F2, whose leading edges are 20.3 microseconds apart. The gating signal and the leading edges of the incident pulses are fed to a coincidence gate where an output signal indicates a valid SSR return. The advent of a 'Framing Pulse Identified' (FPI) signal sets the busy logic and initiates code train timing.

Code train timer

When the presence of a potentially valid code train is established by the FPI, the timing is continued by code train timers associated with one of two identical code registers, the use of two code train timers and code registers enabling code trains containing interleaved pulse positions to be decoded. When the decoding process is complete the code information stored in the register is fed to the plot validation unit.

Garble detector

The garble detector recognises conditions which cause a garble signal to be present at the output of the decode unit. This garble signal occurs concurrently with the signals relevant to mode/code, Special Position Identification (SPI) etc.

Emergency detection

Both civil and military emergency replies are automatically detected and indicated. Detection of a four train emergency causes a Selective Identification Feature (SIF) emergency signal to be generated for display. The civil emergency codes 7700 and 7600 also generate a distinctive signal. All emergency code detectors feed into an integrating circuit which ensures that spurious signals do not give rise to emergency indication.

Height code translation

Height coded replies on mode C are routed via a height transalator to the output. The height translator converts the incoming reflected binary code to binary coded decimal to the nearest 100 ft or 500 ft as dictated by the incoming code.

From the decoder, signals are fed to the plot validation unit.

Plot validation unit

The purpose of this unit is range extraction and plot acceptance, code validation and to perform the plot bearing calculation. It also incorporates on-line test facilities and manual fault finding aids.

Range extraction and plot acceptance

A pulse is received from the decoder for each target reply which allows bracket pulse recognition. The range of each such bracket pulse recognition is stored during the period after the interrogation pulse. During successive periods the stored ranges are compared with those of a number of previous periods, i.e. within a sliding window.

If the number of replies at a particular range in any one of three modes exceeds a leading edge setting, then a plot is initiated in the validation process. As the replies for a particular target move out of the aerial beam, no new replies are stored and the number within the sliding window reduces. As this number falls below a trailing edge setting, the plot validation processes are completed.

Code validation

The code data contained in a reply are stored and compared with the code data received in the next reply on the same mode. If both contain the same information, validation is achieved. If not, the second reply is stored and comparisons continued until validation is achieved when a validation bit is set. The validation process continues until all mode and code related to the plot has been validated or the target falls below the trailing edge criterion.

Plot bearing calculation

The aerial bearing information is fed into the code validation storage when the target exceeds the leading edge criterion and forms part of the data corresponding to the plot. When it falls below the trailing edge criterion, the current aerial bearing and the initial bearing are used to calculate the plot centre bearing, which is then attached to the plot message and held in the validation store. As each plot message is completed it is ready for onward transmission.

Combined primary/secondary radar plot extractor

Previously, primary and secondary radar plot extractors have been described separately, but, where both primary and secondary radar equipments are operated at a radar station it is possible to combine both types of plot extraction within a single equipment.

Within such equipment separate processing channels are provided for primary and secondary radar targets but as the aerials of both systems are mechanically or electronically linked, a common timing unit is sufficient for the generation of digital azimuth and range information. After plot detection the outputs of both channels are fed to a target information process computer for correlation of primary and secondary radar target returns, buffering of the position coordinates of the detected targets and organisation of data transmission.

The data output from the target information process computer is passed to the modems which convert the signals to a suitable form for transmission over telephone lines.

5.3 Radar display

The device whereby the information acquired by the radar equipment is made available to the operator or air traffic controller is known as the radar display.

The form in which the information is presented on the display may take many forms but within the air traffic control environment only three are in common use: the 'A' scope or range display, the plan position indicator and a combined range/azimuth and range/height used with precision approach radar.

At the present time all display systems present their information on a Cathode Ray Tube (CRT). Although development is proceeding on alternative methods such as the plasma and LED displays, these have not yet been developed sufficiently to provide a viable alternative.

The cathode ray tube, which is best known for its service in the television receiver comprises: a thermionic cathode which emits electrons, a series of

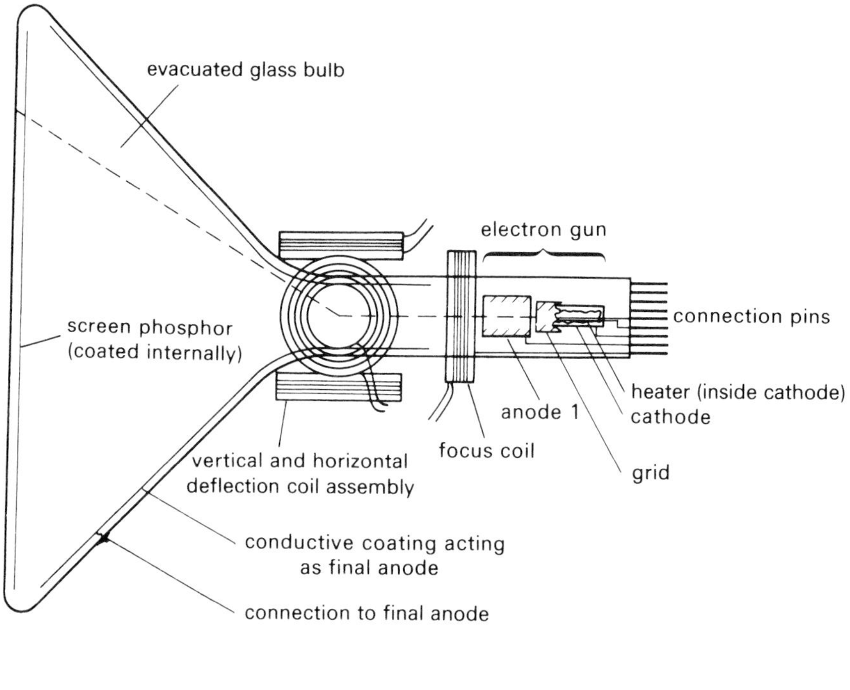

Fig. 38 Construction of an electromagnetic cathode ray tube

electrodes which, by virtue of electrical voltages to which they are subject, focus the electrons into a narrow beam and a screen which illuminates at the point at which it is struck by the electron beam. The focussing and deflection beam may be achieved electrostatically by means of plates placed internally within the CRT or electromagnetically by external coils.

A radar CRT is situated within the display unit and may be of 5 inches to 23 inches diameter depending on the purpose for which it is required.

The display units contain all power supplies for the CRT and generates the deflection voltages necessary to develop the basic scan. The necessary processing and mixing of the incoming signals may also take place within the unit, in which case it is known as 'autonomous display'.

'A' scope display

This display indicates only range information. The spot deflects from the left to right hand side of the CRT face, commencing as the transmitter fires and taking such time to traverse the tube face as an echo takes to return from a target at the maximum range required. The incoming signals are arranged to deflect the beam vertically. Noise within the receiver also causes a vertical deflection and this is normally visible at all points along the trace. CRTs used for 'A' scope display normally have a green fluorescence and, because of the similarity in appearance, noise is usually termed 'grass'. The information renewal rate is identical to the equipment PRF, consequently the CRT phosphor need only be of a type with short persistence of fluorescence.

'A' scope displays are not normally used within the operational environment but are extensively employed by maintenance personnel for setting up and alignment purposes.

PPI display

The requirement to assess the relative position of all targets within the area covered by the radar equipment led to the development of the PPI (Plan Position Indicator) display.

In the most basic form of this display the trace commences normally from the centre of the tube to the edge in a direction corresponding to the direction of the aerial. The brightness of the trace is adjusted to be barely visible and incoming signals are arranged to increase the brightness of the trace. The tube phosphor is selected so that after initial illumination an image remains for a considerable period–often exceeding half a minute.

As the aerial array rotates, the trace (more correctly known as the 'time base') also rotates in sympathy and targets show as short bright arcs in a position corresponding to their range and bearing, their length corresponding to the beamwidth of the aerial.

Although the range of the target corresponds to the distance from centre, this is difficult to assess without further reference. All displays are therefore

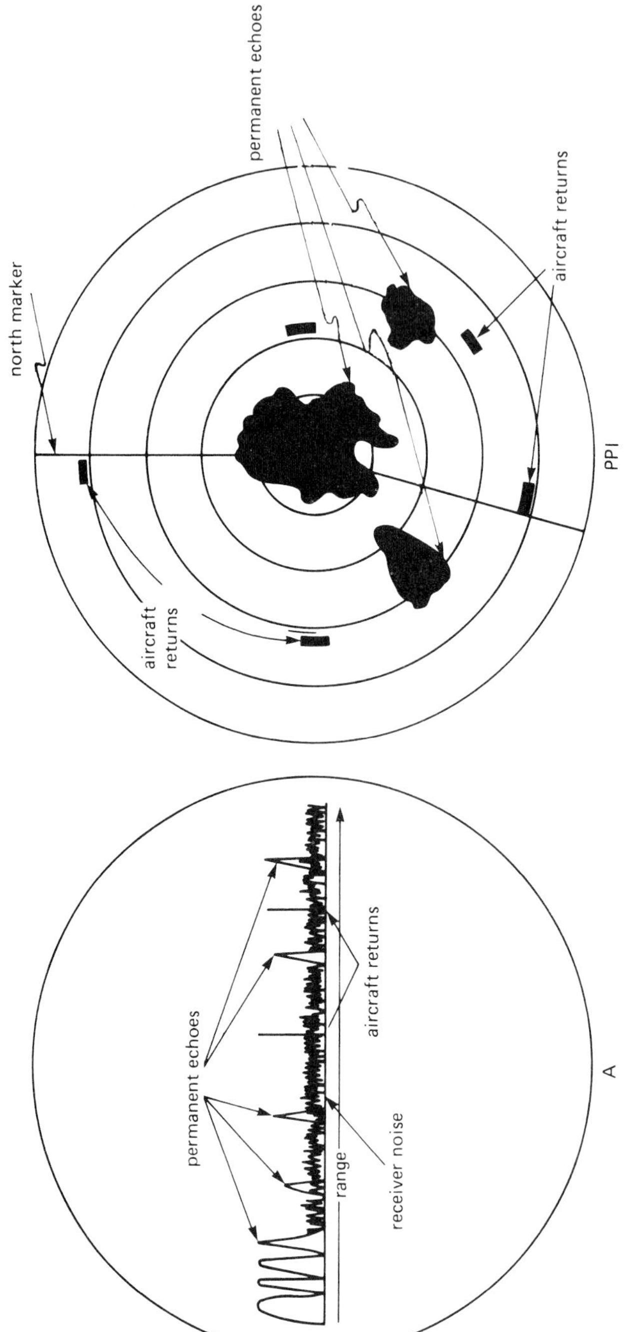

Fig. 39 Type 'A' and PPI radar displays

equipped with a range mark generator. This is an oscillator whose output has been processed to produce short spikes at intervals corresponding to the time taken for an echo to return from a target at specific distances, i.e. 5nm or 10 nm intervals. It is quite common for more than one generator to be fitted which can be switched individually or in combination. The oscillator commences operation at the firing of the transmitter and the output is arranged to 'bright up' the display. The overall result is therefore a series of concentric rings at a spacing corresponding to the range interval selected.

For some purposes it is only necessary to survey a proportion of the area of cover of the radar and for these circumstances arrangements are made to offset the centre of the scan, even to the degree where the origin of the timebase is removed well beyond the edge of the tube. An example of this type of scan is the Distance From Threshold Indicator (DFTI) which is fitted at major UK airports. It comprises a 5 in offset display operating from the airport surveillance radar which displays echoes of aircraft whilst on final approach to the runway.

North marker

Mention has previously been made of a striker attached to the aerial turntable that activates a switch once per revolution for checking the aerial alignment. If activated by applying power, this circuit causes a'bright up' of the trace when the switch is operated by the turntable striker. Conventionally the switch is arranged to operate as the aerial passes magnetic North, and is, in consequence, called the North marker.

MTI range gate

Although an efficient MTI circuit will eliminate permanent echoes leaving aircraft targets visible, nevertheless, a certain degradation of signal returns occurs, particularly at longer ranges. It is therefore normal practice only to switch the MTI into circuit for the first part of the scan and to display non MTI (i.e. raw) video beyond that range. The switching is achieved electronically and the circuit is known as the MTI range gate. The range over which the MTI is operative may be varied by a control on the display unit and is usually set by operator to his own preference.

Fast time constant

If heavy returns from permanent echoes or precipitation are showing on the display it is sometimes advantageous to engage the 'Fast Time Constant' (FTC) circuit. This comprises a passive circuit incorporated after signal detection whose time constant is approximately equal to the pulse length. The action of this circuit is to respond immediately to an incoming signal but rapidly reduce the output level immediately afterwards. The effect of this is

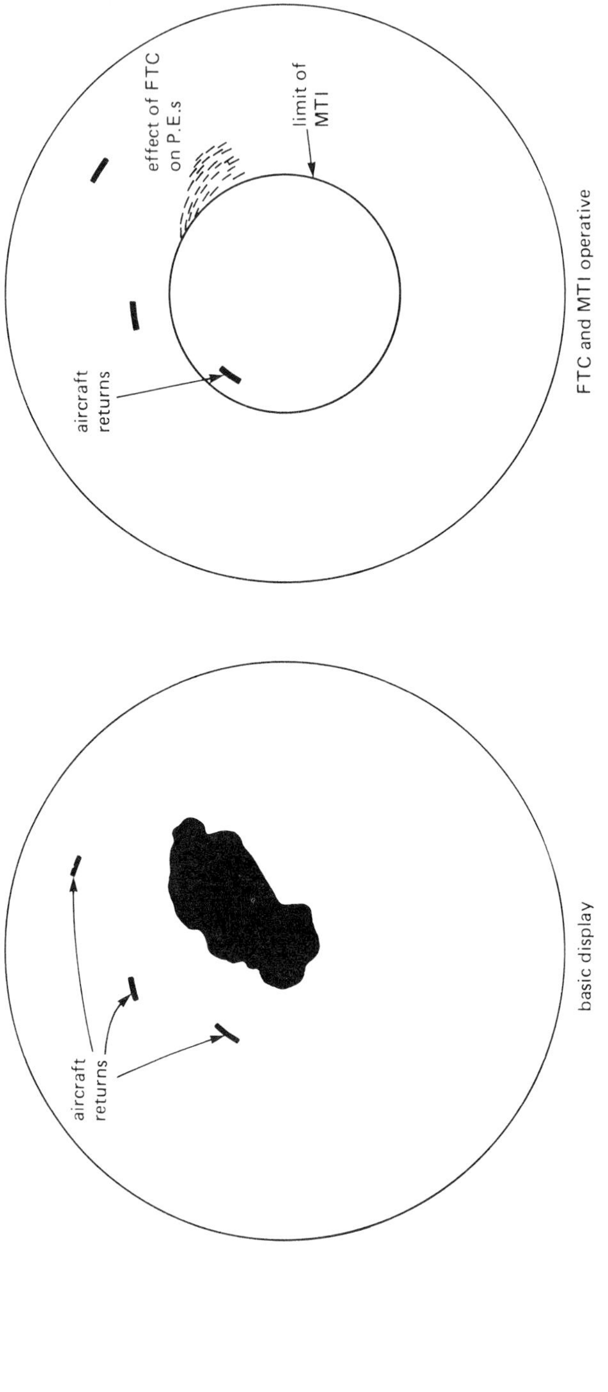

Fig. 40 Effects of FTC and MTI

that returns from aircraft or similar targets are hardly affected but those from large objects such as permanent echoes or weather, are broken up, only showing the edge nearest the origin of the time base.

Video maps

When the permanent echoes have been removed from a PPI display, the air traffic controller or operator has no positional reference other than the range and bearing of the target from the radar head.

In very early radars this problem was overcome by inscribing a map of the area of radar cover on a transparent sheet mounted over the CRT and later solutions included projecting a map on to the tube face by means of a projector mounted above the operator's head. Neither of these solutions proved really satisfactory and both have now given way to electronic video maps. As may be expected, analogue systems have been in use for many years but these are being replaced by digital equipment although it is a reasonable surmise that the two systems will co-exist for many years.

The analogue video mapping equipment comprises a high definition, high intensity CRT, a lens system, a transparency holder and a photo electron multiplier tube.

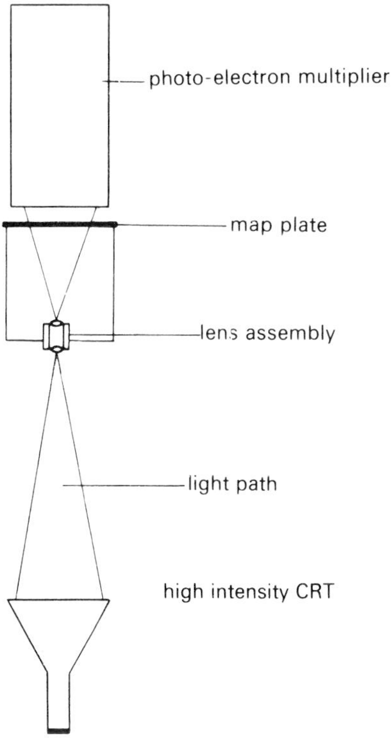

Fig. 41 Analogue video map

The CRT is usually mounted vertically, face upwards in the lower part of the cabinet and it displays a bright unmodulated time base synchronised both in time and position with that on the PPI displays. Above the CRT is a lens system which projects the displayed time base on to a photographic plate of a map, centred on radar head or the area of radar cover. This negative is surmounted by the photo-electron multiplier which is activated by the light from the projected CRT timebase passing through the clear sections of the map negative. The output provided by the photo electron multiplier is a video signal which, after suitable processing, is fed to the PPI displays where the resultant is a reproduction of the map plate.

In many installations, digital equipment is now in service. In this, all the information to be displayed is held in a memory circuit from which it may be called as and when required.

The output sequence starts during the interscan period when the aerial turning signals, which define the bearing of the aerial and the direction of rotation, are read into a separate store, used to search for a corresponding set of bearing data in the main memory circuit, and when found, output the address onto a further output store.

During the scan period each instruction in the output store is read from the main store and decoded at the correct time by comparing with an internal range counter. The counter is zeroed at the start of each scan and, when the count corresponds to the first instruction of that bearing, a signal output is sent to the radar display unit. The next instruction is then read and the process continues until all the instructions for that bearing have been completed.

At the completion of the scan, the interscan period is entered and the equipment decides, from any change in the bearing information, whether the information in the output store should be replaced by a new set of bearing data.

Although the image generated on the PPI tube is termed a map it bears little resemblance to maps as used by the general public, the information shown being aeronautical in character such as airways, reporting points, danger areas, extended centre lines of runways, etc.

Selective MTI

At one airport in the UK it was noticed that most of the PEs occurred within certain well-defined areas and the use of the MTI range caused signal degradation over considerable areas where MTI was unnecessary. It then occurred to the staff that if a special video map plate could be manufactured outlining the areas of permanent echo, the output from the video map generator could be used to switch the MTI circuits in a similar fashion to the range gate, thus only operating the MTI circuitry where it was required. This was tried and proved to be successful.

Since then the idea has been further developed by the manufacturers of the video map equipment.

Interscan

There is a considerable time lapse, in radar terms, between the end of one scan and the commencement of the next. This may be typically in the order of 1250 microseconds for a radar with a range of sixty miles and a PRF of 500 Hz. During this period the video and spot deflection circuits are inactive and may be used to accept inputs from other sources. These may include the display of an 'Interscan' line or alpha numerics such as a clock.

The 'Interscan' line is a straight line which may be produced on the CRT at any length, bearing or position. The controls for this line are accurately calibrated, thus it may be used for measurement or the marking of a predetermined point or track for the convenience of the air traffic control officer or radar operator.

Due to the persistence of the CRT, the line (or other 'Interscan' images) appears simultaneously with and is indistinguishable from images produced during the scan period.

Direct view storage tube displays

Although modern cathode ray tube phophors, particularly when combined with scan conversion and similar techniques, give more than sufficient light intensity to be used in conditions of high ambient light, this is a relatively modern development.

In the past, PPI CRTs were confined to areas of low ambient light level; and where viewing in full daylight conditions was necessary, such as in ATC visual control rooms, other means had to be developed. A widely used answer to this problem was the use of the Direct View Storage Tube (DVST) Display, which still remains in service in many parts of the world.

The DVST has two electron guns which are mounted co-axially to facilitate uniform tube face illumination and minimum picture distortion. One gun writes the picture as a charge pattern on a metallic storage surface using high velocity electrons. This mesh is flood illuminated with low velocity electrons from the second gun. Where a pattern has been written, these electrons are accelerated through the mesh in parallel streams continuously to energise a high output phospor on the tube face. Instant erasure is possible and is initiated either by the operator or automatically on changing range. Erasure is also carried out as a continuous process to remove decrementally the stored information and to limit the charge build up on the storage mesh. The auto-erase pulses which produce slow total erasures are called 'background erase'.

They are supplemented by a system known as 'highlight erase' which causes rapid partial erasure. The two types of continuous erasure combine to give the initial flash and decaying trial on moving targets which characterise the conventional fluoride radar display.

With the exception of the additional controls associated with DVST these displays have the same facilities and similar operational characteristics to normal fluoride displays.

The DVST will provide a picture which is several hundred times brighter than the conventional fluoride tube and may therefore be used under all ambient lighting conditions, even direct sunlight.

Distance from threshold indicator

The limit on the number of aircraft movements in and out of an airport is very often a function of the utilisation rate of the runways. To assist the controller achieve maximum traffic flow a specialised PPI display has been developed which shows only the approach path to the operational runway. By use of this equipment he is constantly in the position of knowing the exact position of aircraft on the approach path and in consequence is able to judge when outbound aircraft may be safetly released for take-off.

The equipment comprises an offset PPI display utilising a 5 in diameter DVST. Up to six off-centre positions may be set into the system, these representing alternate ends of three different runways. Selection is by rotating the cursor surrounding the tube face which is fitted with click-stops. When a runway has been selected by this method, the threshold appears at the zero range point of the display regardless of the actual physical position of the radar head. Should a short range PPI presentation be required, for example, for monitoring overshoots, this can be provided by the depression of a switch.

The graticule on the cursor is engraved with two parallel lines marked at nautical mile intervals which bracket the runway centre line and in addition a range mark generator is also available.

The equipment may be fed from any radar system giving the appropriate cover and will also accept video map or external range mark video.

Scan conversion

The quest for brighter radar displays also led to the development of the scan conversion technique. At its simplest this takes the form of a closed circuit television camera viewing a standard fluoride radar display. The operator or air traffic controller views a TV monitor which, due to the high information refresh rate, is sufficiently bright to be viewed in normal intensities of room lighting.

In more common use, however, is equipment which uses a special scan converter cathode ray tube. This tube is dual-ended and has at one end an electron gun, focussing and PPI deflection assemblies which fires a high velocity electron beam at a fine wire target mesh. This is covered with a layer of insulating material on the side facing the electron gun only. The resultant secondary-emission electrons are attracted to a collector mesh maintained at

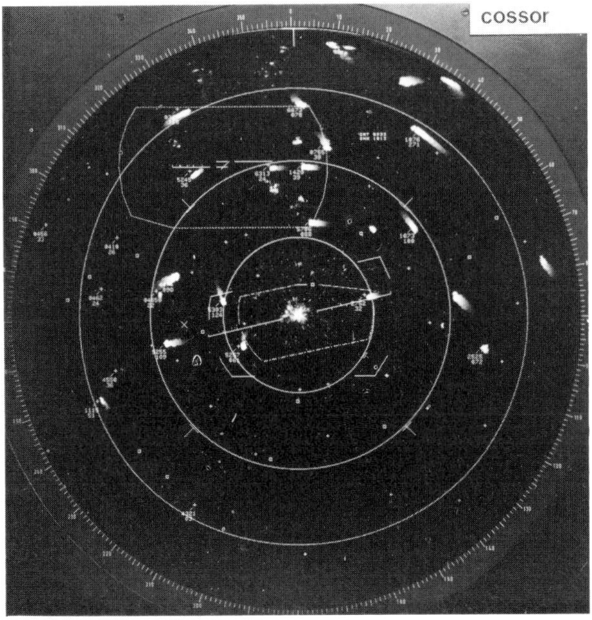

Plate 33 This photograph of the Cossor CDT 3000 display shows the definition possible on a 1500-line scan-converted display. Indicated are raw radar and SSR targets, range rings and video map. *(Photo: Cossor)*

a suitable potential placed between the gun and the target. The target mesh is therefore left positively charged at a potential depending on the degree of secondary emission, the effective target capacitance and the intensity of the electron beam. This is known as the 'write' side of the tube.

The stored information is read out by firing a low-velocity 'read' beam at the target mesh from the opposite end of the tube and by decelerating the beam to near-zero velocity when it reaches the target mesh. The space-charge potential existing on the storage surface determines the proportion of read-beam electrons that pass through the mesh and those that are reflected back towards the read gun where there is an electron multiplier which provides signal gain. The majority of the electrons passing through the target mesh strike the collector mesh but a small percentage reaches the storage surface and slowly neutralises the positive charges stored there.

By use of this system it is therefore possible to 'write in' a standard PPI picture and 'read out' a CCTV picture. For air traffic control purposes, however, the definition capability of the normal 625 line television standard is inadequate and it has been found necessary to increase the number of scanning lines to 1000. In addition the scanning format has been changed from the normal left to right with a rapid flyback to an alternating left to right, right to left. When displaying plot extracted secondary radar it is frequently necessary to include aircraft labels or other information. At these times, the read scan is stopped by the display processor which writes the required legend and then permits the read scan to continue.

Due to the techniques used in scan conversion, the operator or air traffic controller has no control over the range displayed or picture offset, the only adjustments that can be made to the display being brilliance, focus and gain. This is of little disadvantage, however, within the control centre environment for the normal mode of operation is for one radar suite to be allocated to a single control area and in consequence there is little need for other than a standard picture. If for some operational reason observation of another area is necessary, the display can be connected to the output of another scan converter.

Computer generated displays

With the introduction of plot extracted radar equipments, a movement away from traditional PPI and scan converted displays could be initiated. The signals from a plot extracted radar consist of a series of digital words, each corresponding to the position of, and SSR derived data from, a single radar return. This is in suitable format for feeding to a small general purpose computer, known as the display processor, which in accordance with a predetermined program, will control the display unit spot deflection circuits in such a way as to write a representation of a PPI display. Targets are indicated by mathematical symbols and SSR derived data (identification and mode C height response) is displayed in alpha-numeric characters adjacent to the target symbol.

The mode of operation is that the incoming data is placed in the computer store, from which it is read in a sequence determined by the program. As the action of reading from store does not erase the contents, the information retrieval rate is independent of the information renewal rate from the radar and in consequence can be sufficiently high to ensure a bright and flicker-free picture. Further additions to the program can generate a video map presentation and, interfacing with an additional, suitably programmed computer, can arrange SSR code/callsign conversion when that information is available.

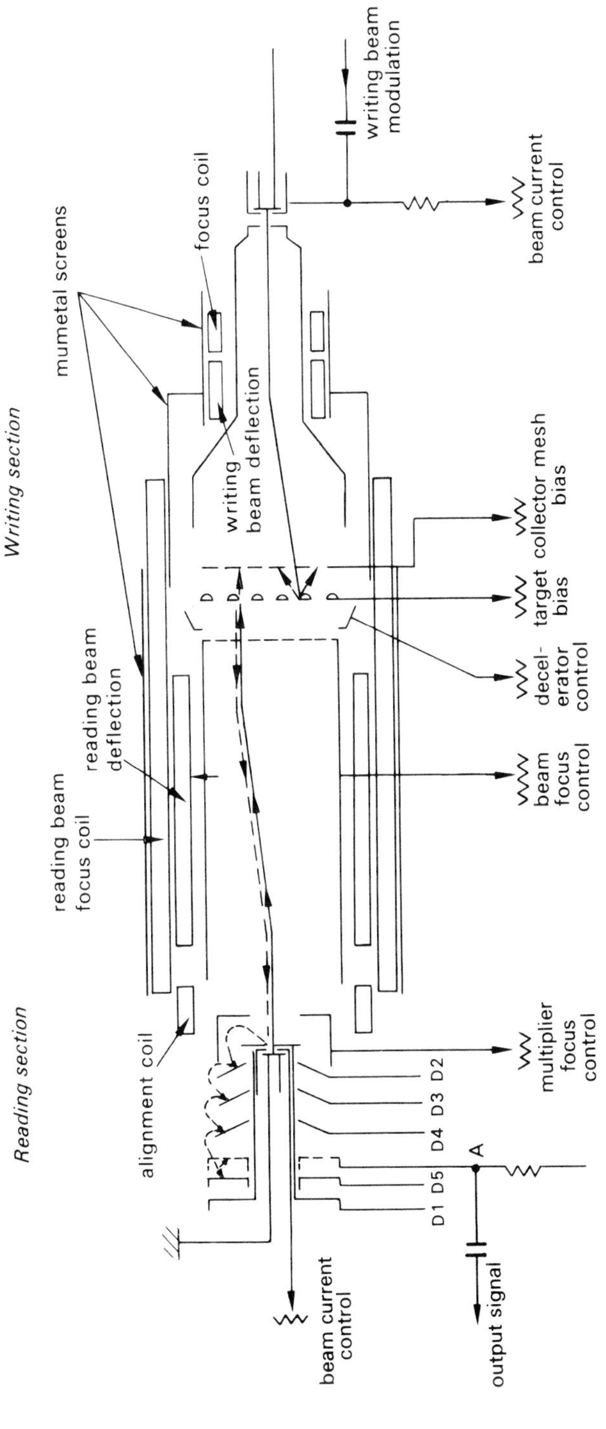

Fig. 42 General schematic of a scan converter tube

5.4 Radar data processing systems

The development of both primary and secondary plot extraction systems has made possible the integration of radar data from different sources in a way which was inconceivable only a few years ago. In the past, the signals received from each radar station were individually processed and the air traffic controller could only view signals from one radar at any given time. The incorporation of a major computing installation not only integrates the output signals from all available radars into a composite system to ensure that the best possible picture is displayed at all times, but also facilitates the display of non-radar and computed data. This may include such additional data as flight plan information, predicted flight path and arrival lists depicting aircraft which have flight plans within the system and are within a parameter time of their ETA at the centre boundary.

An IBM 9020D computer was installed at London Air Traffic Control Centre in the early 1970s; it was used initially for flight data processing (FDPS) and later, when the necessary system development had been achieved, for radar data processing (RDPS).

By the late 1980s it became necessary to replace the 9020D by the more modern IBM 4381, more commonly referred to as the National Air Space Host Computer System or NAS-HCS.

For both operational and economic reasons, the 9020D software was rehosted onto the IBM 4381 and consequently the changeover was transparent to operational staff.

Multi-radar data processing

In a conventional system, an air traffic controller can only use the signal returns from one radar even though it may not be ideal. The RDPS collects the digitised data from all available radars and constructs a composite (mosaic) picture covering the whole of the FIR airspace. Each controller may then select the portion of the total picture relevant to his area of responsibility.

To achieve this, the airspace is divided into a grid of 16 nm squares forming columns of air from the ground upwards called Radar Sort Boxes (RSB). Each of these has up to four radars allocated which can provide service in that area. The radar giving the best cover is nominated as 'preferred', the next best 'supplementary' and the remainder form a reserve. Whenever the information from the 'preferred' radar is satisfactory it will be used but if an expected response from a tracked aircraft is not received which is available from the 'supplementary' radar then this will be processed for display and tracking purposes. Should the 'preferred' radar fail, then the 'supplementary' radar can be upgraded either manually or automatically to 'preferred' and a reserve nominated as 'supplementary'. In exceptional circumstances two radars can be designated 'preferred' in the same radar sort box, but this considerably increases computer processing time and has other disadvantages on the display.

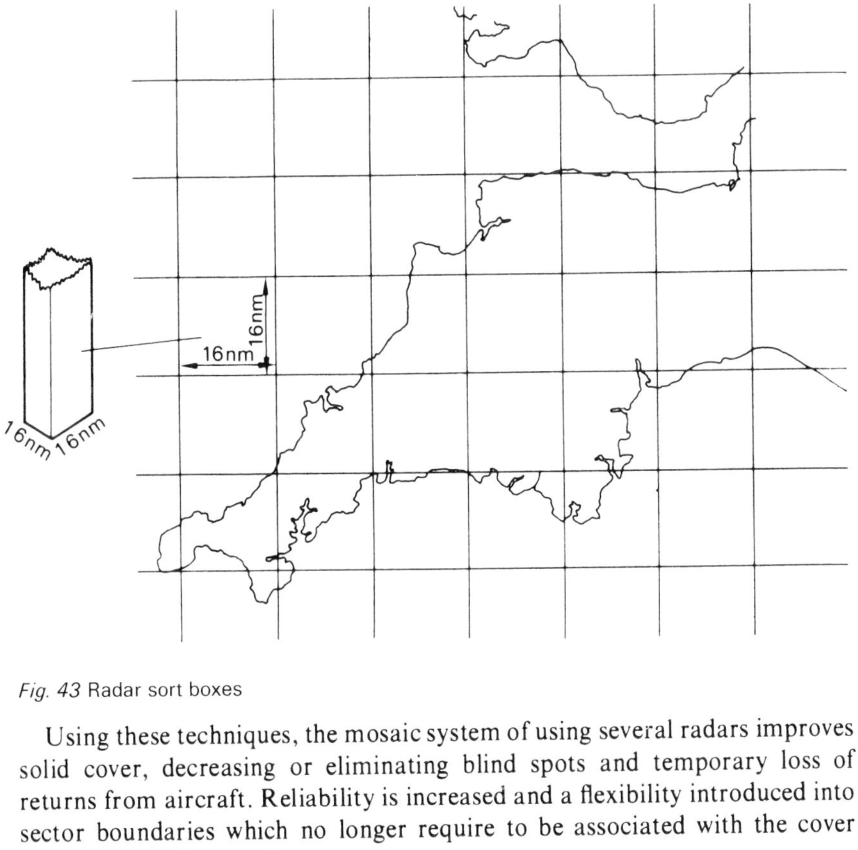

Fig. 43 Radar sort boxes

Using these techniques, the mosaic system of using several radars improves solid cover, decreasing or eliminating blind spots and temporary loss of returns from aircraft. Reliability is increased and a flexibility introduced into sector boundaries which no longer require to be associated with the cover provided by a single radar.

When the RDPS receives plot extracted radar targets, certain positional corrections, such as for slant range, are applied. The computer program then checks to see whether these can be linked to a track already established within the system. Those which can are passed on for display and to the automatic tracking program for further processing. The remainder are passed for display only.

Automatic tracking

The primary objective of automatic tracking is to maintain the unambiguous tracking of aircraft targets. To achieve this it is necessary to bring together both the radar information and the flight plan to create a common flight data base upon which each subsystem (FDP and RDP) can draw in the execution of its tasks. The past movements of targets are combined with the flight plan intentions to enable the progress of flights to be monitored and predictions to be made for the future.

When a target has been correlated with a track, the target velocity is calculated and this, together with the flight plan intention allows the prediction of a search area where the next target may be found.

By this means the computer maintains an automatic correlation between track and target even if there are other aircraft within the vicinity responding in the same SSR code. It is this facility which earns the system the title 'automatic' and which differentiates it from manual tracking in which teams of trackers have to maintain visual identity of each target and, by rolling ball or joystick, identify it with the track to which it belongs. Automatic tracking can be maintained in the case of correlated aircraft with a high degree of confidence but non-correlated SSR and primary returns are, under certain circumstances, subject to track swapping and in the case of primary returns, weather clutter confusion.

There are three dynamic modes of tracking: FLAT, FREE and COAST. The mode in use may be identified by a symbol displayed adjacent to and often superimposed upon the target symbol.

The FLAT (Flight pLan Aided Tracking) mode is the highest level of tracking and makes use of both radar and flight plan data. The target is related to the route segment being flown and any lateral deviation noted. If this is within a predetermined parameter, the aircraft is considered to be conforming reasonably to its flight plan. Longitudinal variations will result in the forward fix estimate being updated and if this is significant the revised estimate will be relayed to interested sectors.

When a target has deviated from its planned route or no route data are associated with the target, tracking is reverted to a second level, known as FREE. Longitudinal checks cannot be made and the prediction of future movement is based on the extrapolation of past target data.

On occasions, when a target is being tracked in either FLAT or FREE mode, signals may cease for some reason. Under these circumstances tracking will continue in the FLAT COAST or FREE COAST mode by prediction using the last known velocity assisted in the former case by flight plan route data. This will continue until signals are again received or until a predetermined time limit has expired when the track will be automatically deleted.

Data display on the radar screen

As the radar screen is the centre of the air traffic controller's attention, it is a logical place for the display of additional data which will be of assistance in the control of air traffic.

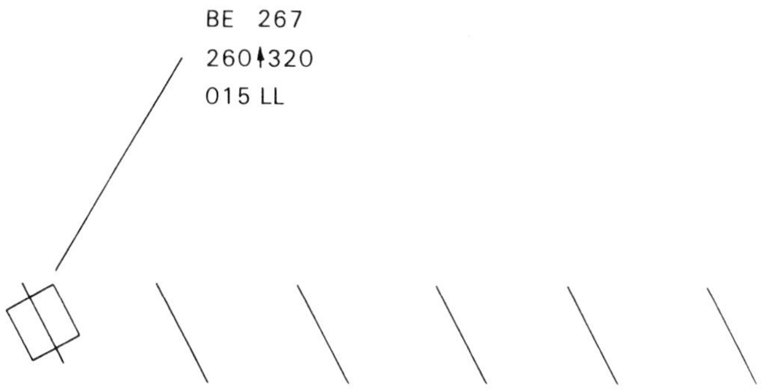

Fig. 44 A tracked return with full data block. The position symbol indicates that the aircraft being tracked in FLAT mode and the track history is shown by the five diagonal lines to the right. The data block informs that the aircraft, callsign BE 267, is at flight level 260 climbing to an assigned level of 320, en-route for London Airport (LL). 015 is a computer identification number for the flight.

The displayable data can be divided into three broad categories:

(a) Essential data such as target returns, data blocks and maps. These may be selected on a keyboard.
(b) Information to assist in preparing messages for the computer i.e. preview area for building up the message (scribble line) and the computer originated area (for error details). These are displayed only for the duration of the operation.
(c) Lists for reference. These are selected from the keyboard and only displayed when required.

Essential data display

The position of the target is indicated by a symbol which by its form indicates whether the return is from primary or secondary radar, tracked or untracked. Coincident with this, or closely adjacent, is a further symbol which indicates the type of tracking in use. The history of the track is represented by synthetic representation of the trail, or afterglow, of a traditional raw radar in the form of up to five target symbols displayed in five variable brightness categories. The number of history symbols displayed is selected on the operator's keyboard. Adjacent to the target symbol is situated up to three lines of target information, known as a data block, which could include such data as aircraft call sign, flight level, flight level changes in progress, destination, computer identification of flight plan, etc. If any emergency codes are received, these will be displayed in the data block with flashing symbols.

Maps displayed on the radar viewing unit are generated by a computer program. A number of maps are available, each depicting a different point of view such as: beacons and reporting points; Airways; SST routes; danger areas and coast outlines. Each sector has a map area suitable for its own needs which is available in five levels of information. By use of various selections of these levels of information, the controller may select the complexity of mapping information which he requires.

Weather returns received by the incoming radars can be selected on the keyboard. Such returns are indicated on the display by a series of hash lines.

Computer messages

To assist a controller in compiling messages for input to the computer a message preview area is available. As he enters the message using the keyboard, so the relevant letters or numbers appear on a 'scribble line' on the display. When the message is complete, the enter key is pressed and the message disappears from the display. In a similar fashion, computer replies to the controller may also be displayed.

Lists

A number of lists are available for display. These include arrival lists, altitude limit lists, SSR code select lists, etc. The list required is selected by keyboard action and is displayed in the least used area of the screen.

The rolling ball

The rolling ball is essentially a positional entry device. It controls the position of an electronic marker on the viewing unit screen which is used to identify any indication on the screen which is to be the subject of a computer destined message.

Future developments

The versatility provided by a major computer installation allows additional facilities to be added as and when suitable software becomes available. As at early 1992, for example, work was proceeding on the installation of a short term conflict alert facility. This is a radar based tactical conflict detection system which functions completely automatically thus imposing no extra workload upon the controller. In effect, it looks ahead two to three minutes and provides just sufficient time for remedial action to be taken without giving longer range warning which would have been dealt with in the normal course of events. The intention is primarily for a 'last ditch' safety facility to avert trouble when the controller has missed a situation due to heavy workload.

Other possible developments include:

(a) *Flight plan probe*, which is meant as a conflict probe at cruising levels on overflying aircraft.

(b) *Flight level allocation*, which will assist in arranging flight levels on outbound and overflying aircraft so that they are in a safe pattern before exiting to adjacent centres.

(c) *Flow control*, which will give warning of capacity overload, and suggest the best course of action. Additionally, this should facilitate control of outbound traffic patterns taking into account the limitations imposed by adjacent centres.

Section 6
Hyperbolic navigation systems

6.1 Decca Navigator

The Decca Navigator system was developed towards the end of the Second World War. Originally designed for maritime use this aid has subsequently been further developed for use with both aircraft and hovercraft.

Operating on a combination of frequencies on the LF band, Decca Navigator is a continuous wave hyperbolic navigation system which is useable up to about two hundred and forty miles beyond which the combination of ground and sky waves causes degradation in accuracy.

The principle of operation can best be described by initially considering two transmitting stations spaced by 15 km, radiating synchronised carriers on a frequency of 100 kHz, i.e. 3000 m wavelength. On the line between the stations there will be ten places at which the signals will be in phase. If a map is then drawn showing all points at which the two signals are in phase it will be found that a series of hyperbolae are formed with the two stations as the foci.

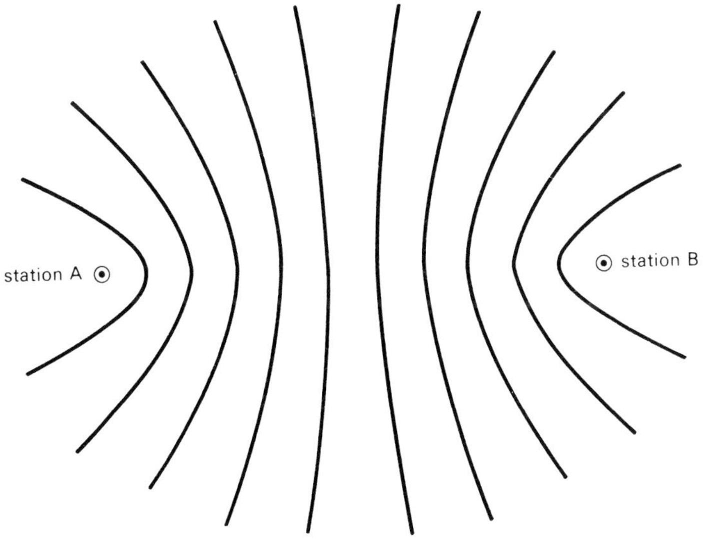

Fig. 45 Hyperbolae of in-phase signals

Furthermore, at any point between those hyperbolae the phase difference between the signals from the two stations will remain constant. By counting from one transmitting station the number of times the signals have been in phase and adding the phase difference between the two signals, it is therefore possible, to calculate the relative ranges of the two stations from any location within range of both stations.

If a third transmitting station is added to the chain, two further families of hyperbolae may be added to the map and an accurate position 'fix' obtained by comparing signals from each pair of stations.

In practice, if such a system were tried there would be many difficulties, not the least being the identification of individual signals. The Decca Navigator system overcomes these difficulties by radiating a different frequency from each station, each of these being a multiple of a common frequency.

A Decca Navigator chain consists of four stations: a master, to which the others are synchronised, and three slaves designated red, green and purple. Their frequencies bear the following relationship–master: 6f, red: 8f, green: 9f and purple: 5f, where f is in the order of 14 kHz and varies from chain to chain.

The master station radiates a crystal controlled transmission at a power of approximately 2 kW to a 300 ft vertical aerial. The slave stations are similar except that their equipment design is such that their carrier frequencies may

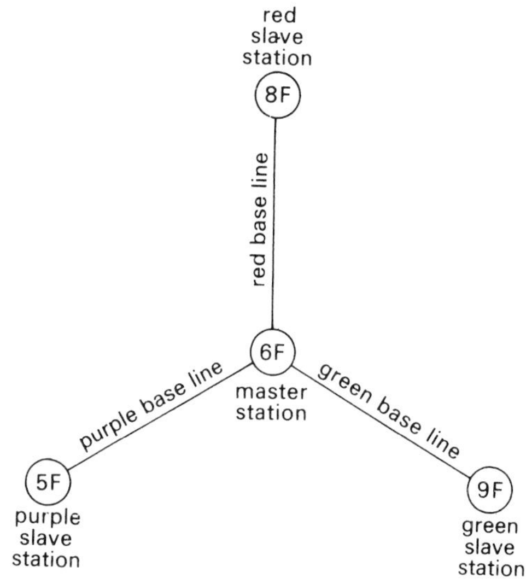

Fig. 46 Layout of a Decca chain

be synchronised to the master transmission. By this means each station radiates a stable continuous wave transmission which bears a constant frequency relationship to the other three stations.

The phase comparison is made in the receiver by frequency, multiplying the incoming signal from each station by various factors such that comparison may be made at the lowest common multiple of frequencies being compared. Comparison is always between master and slave.

Thus, considering a practical case where the master station is operating on 85 kHz, the operational and comparison frequencies may be tabulated as follows:

Station	Frequency	Multiplication	Comparison frequency
master (6f)	85.00 kHz	4	(24f) 340 kHz
red slave (8f)	113.30 kHz	3	(24f)
master (6f)	85.00 kHz	3	(18f) 255 kHz
green slave (9f)	127.50 kHz	2	(18f)
master (6f)	85.00 kHz	5	(30f) 425 kHz
purple slave (5f)	70.83 kHz	6	(30f)

Decca lane

In the introduction to this chapter it was described how the phase of radiated signals from the two transmitting stations was coincident at half-wavelength intervals along the base line. As within the Decca system, comparison is not made at operating frequency, coincidence is at multiples of half wavelengths at comparison frequency. The space between adjacent hyperbolae of coincidence is called a lane and the position of the receiving station within the lane is indicated by a continuously integrating phase meter known as a Decometer. This indicates total cycles and fractions of cycles and is accurate to about one fiftieth of a lane, corresponding to about 5 m along the baseline.

Lane identification

In the original conception of Decca, maritime applications were the main consideration and it was felt that the continuous nature of the transmission in conjunction with the relatively slow movement of the vehicle would make lane identification unnecessary as any dropout such as power interruption etc., would not be of sufficient duration to cause loss of lane. This assumption was found to be invalid, particularly for aircraft installations. A method of lane identification had therefore to be developed and this was introduced in 1948.

The lane identification system uses a comparison of the fundamental

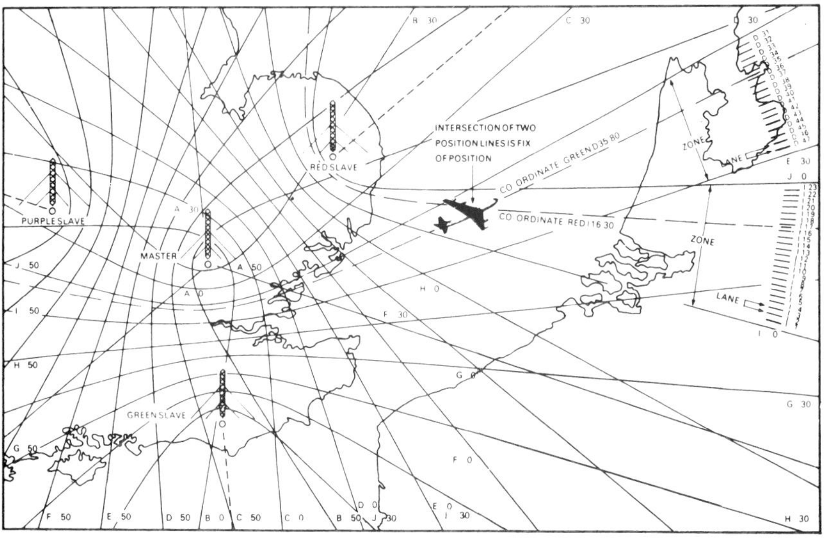

Fig. 47 The layout of the English chain showing how a fix is obtained from the readings of two decometer indicators.

frequency (f) for each of the phase comparison systems for half a second once per minute, producing lanes based on the 14 kHz fundamental, about seven miles apart on the baseline. The frequency 'f' is derived by beating two carrier frequencies together in the receiver. During the lane identification periods only one slave transmits with frequencies of 8f and 9f.

Decca flight log

To simplify the presentation of navigational information in the aircraft cockpit, the flight log was developed. In this type of display the position of the aircraft is represented by a pen which traverses a moving map, indicating both present position and track history. The pen is driven across the map by mechanical gearing and map movement permits movement at right angles.

In the earliest models the Decca lanes were represented by straight lines and in consequence the maps were heavily distorted. In recent years, the processing ability of modern logic circuitry has facilitated considerable improvement in this aspect and the map distortions have been considerably reduced.

Decca Flight Log is no longer manufactured but examples remain in use in limited quantities. The modern tendency is for the information from navigation equipment to be coupled directly to an integrated navigation system such as the RNS 5000 described in Chapter 7.3.

The Mark 15 system

As with all other Decca Navigator equipment, the Mark 15 operates in conjunction with ground-based transmitter chains. However, in the case of earlier equipment, position in, and passage through, the position line lattice was indicated on decometers, the readings of which were used to set up the flight log. In the Mark 15 system, the decometers are eliminated and the output is now in terms of Decca zones (groups of lanes). The only indicator other than the flight log is a simple zone identification meter. Position fixing within the zones is automatic after referencing and requires no further intervention by the pilot.

A further innovation incorporated in the Mark 15 is the provision of a run/fix facility which makes chart changing a far simpler task than with previous equipment. This is accomplished in the computer which stores the Decca position line coordinates during the chart change. On selecting the new chart the pilot is required only to place the pen on a marked reference point from where the indication automatically takes up the correct position.

During the referencing process, selection of 'ref' on the control unit produces an artificial zero for the computer red, green and purple phase comparison circuits. This represents a chart position where the zone boundary position lines intersect. Manually placing the pen on the nearest intersection with 'ref' selected, brings the flight log into agreement with the receiver output. Subsequent switching to the normal operating position brings the pen to the correct zone fraction setting.

The Decca Mark 32 system

In contrast to the Decca equipment previously described, the Mark 32 is designed to operate solely with the Racal Avionics Navigation Management System and, in consequence, does not require any discrete display unit. The receiver is tuned remotely by the navigation computer and the output is in such a form that it can be processed by the computer for navigation function and display.

The Mark 32 receiver

By phase comparison of the master and slave signals, the receiver calculates the Decca Lane fraction for the red, green and purple fractions, which are output to the navigation computer.

From the navigation computer, the receiver receives rate-aiding velocities which ensure that the lane fractions remain accurate during signal discontinuities, such as Lane Identification (LI) periods.

LI transmissions cause the receiver to switch to an alternative mode in which the multipulse LI signal is used to permit master-slave phase comparisons at a different frequency from those used for the normal pattern.

LI values are calculated as lane numbers and fractions and fed separately to the computer. The LI is not permitted to alter the Mark 32 normal lane pattern values.

Zone identification signals are not received by the Mark 32 equipment as adequate position checking facilities are provided by the alternative navigation plots of the Navigation Management System.

Automatic receiver calibration is carried out at intervals throughout normal operation. This is achieved by injecting signals into all four receiver channels. If this does not result in a zero phase difference, appropriate correction is applied to the relevant lane fraction output.

This task is carried out at 5-minute intervals, except during the first 10 minutes after switch-on or on lane change, during which it is performed at 1-minute intervals.

Automatic chain selection

The receiver is capable of receiving any Decca chain in the world and tuning is accomplished by the RNAV-2 navigation computer.

During initialisation of the system the computer scans its chain data store and determines which Decca chains lie within 250 miles. It then constructs a priority list of chains, headed by the nearest. The remainder are then set in order according to pattern geometry. A command is then issued to the receiver to tune to the Decca chain at the top of the list.

This priority list is continually revised during flight, but in order to reach the top of the list, a chain must have better pattern geometry than that currently in use and, furthermore, satisfy certain other criteria. Chain change is delayed until the 'new chain' geometry is some 14% better than that currently in use. This provides a hysteresis which prevents repeated chain changes if the aircraft is flying along a track where both chains provide similar pattern geometry.

If desired, however, the operator may manually select a chain, thus overriding the automatic chain procedure.

The Mark 32 receiver in association with the RNAV-2 computer thus brings the Decca system in line with the current system based navigation systems, removing the workload associated with previous Decca airborne systems.

In normal circumstances the operation is entirely automatic, but should a malfunction occur, or a situation arise which requires a decision from the operator, warning signals are initiated.

Accuracy

With the Decca Navigator systems the only factors which can affect the accuracy of the chain are the speed of propagation of the signals and the accuracy to which the frequency and phase of the slave stations is maintained.

The speed of electromagnetic radiation is a physical constant and the frequency and phase of the radiated signals are maintained to sufficient accuracy to allow a fixing accuracy better than 100 m under good conditions close to the station. At longer ranges the sky wave can interfere with the ground wave causing inaccuracies of reading, the range at which this becomes intolerable being typically two hundred and forty miles. Further inaccuracies are caused at longer ranges when the 'cut' between the lanes being used is shallow. Under these circumstances a fix may be in error by several kilometres.

6.2 The Loran systems

Loran-A

Loran was originally developed in the Second World War with the intention of producing a long range hyperbolic navigation system which did not suffer from the sky wave contamination typical of continuous wave systems such as Decca Navigator. To indicate this purpose, the name selected was an abbreviation of LOng RAnge Navigation.

In any long range system, both ground and sky waves must be present at the receiver, however, the sky wave must necessarily have travelled a greater distance than the ground wave and consequently the two signals may be recognised readily by their relative times of arrival. To facilitate this identification, a pulse system was proposed which used a cathode ray oscilloscope as an indicator in the aircraft receiver. The most suitable operational frequencies for any navigational aid which utilises ground wave propagation are within the LF waveband but at the time of development of this aid it was not considered feasible to generate sufficiently accurate pulses at these frequencies. In consequence the system was placed slightly above the high frequency end of the medium wave broadcast band. This selection severely limited the operational range and doubtless the system would have become obsolete much earlier had it not been for the relatively large numbers of aircraft and ships equipped and the late development of its successor Loran-C. At its peak about twenty-five Loran-A chains were in operation but in recent years these have been gradually closed down, the only station within the UK, Mangersta on the Hebridean Island of Lewis, being withdrawn from service in 1977. All Loran-A chains have now closed.

Loran-C

In the development of Loran-C, a complete break was made from the frequencies previously used, however, in common with the earlier system pulse techniques were again used in conjunction with a hyperbolic system. The operational frequency of all chains is 100 kHz which enables an operational range in excess of one thousand miles to be achieved. In consequence master and slave stations could be separated by up to eight hundred miles.

The Loran-C transmitter

Each Loran-C transmitter operates on a frequency of 100 kHz, generating a

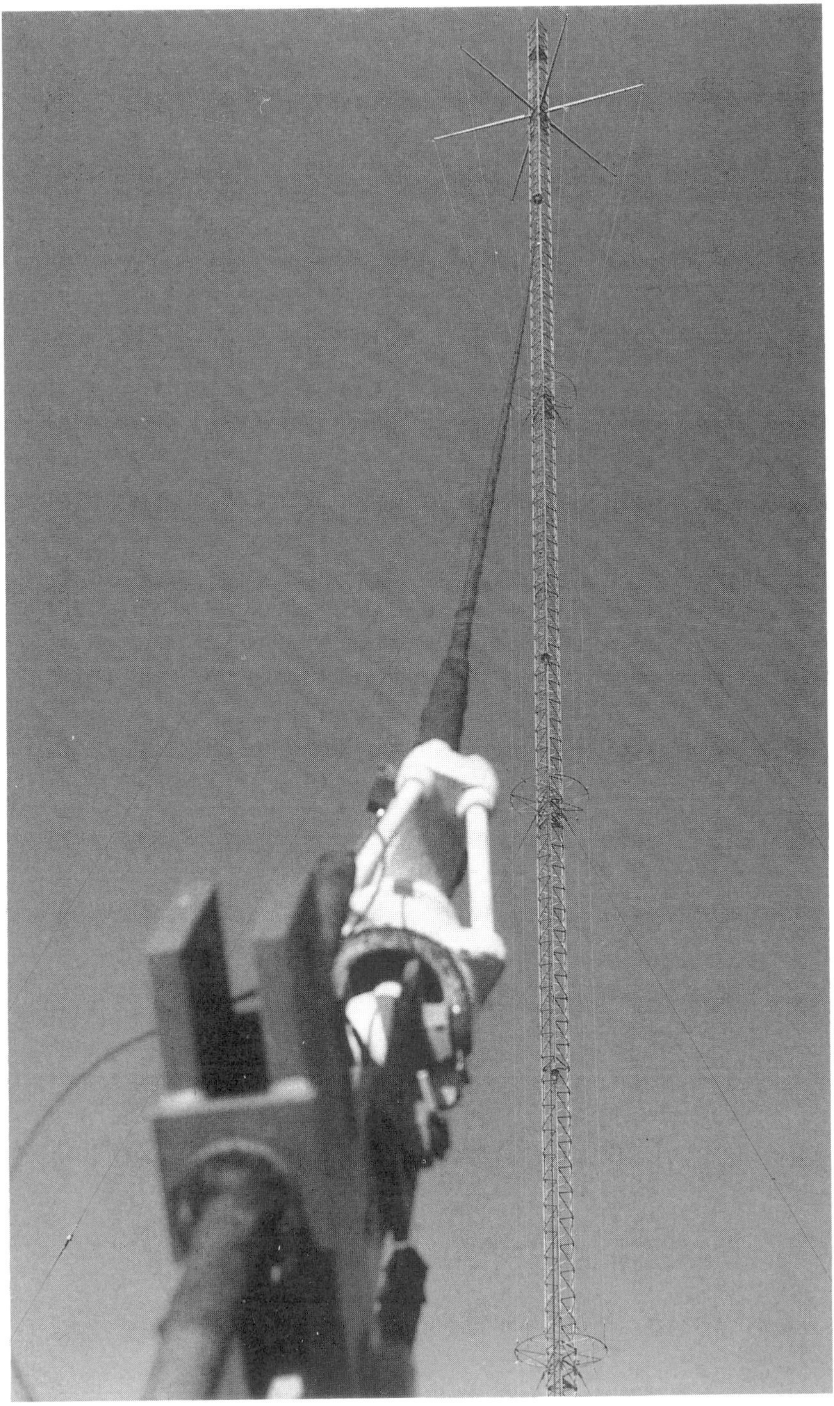

Plate 34 A Decca navigator station transmitting mast. *(Photo: Decca)*

Plate 35 The Decca Navigator Mk 32 receiver with the R-Nav2 computer installed in a helicopter. *(Photo: Decca)*

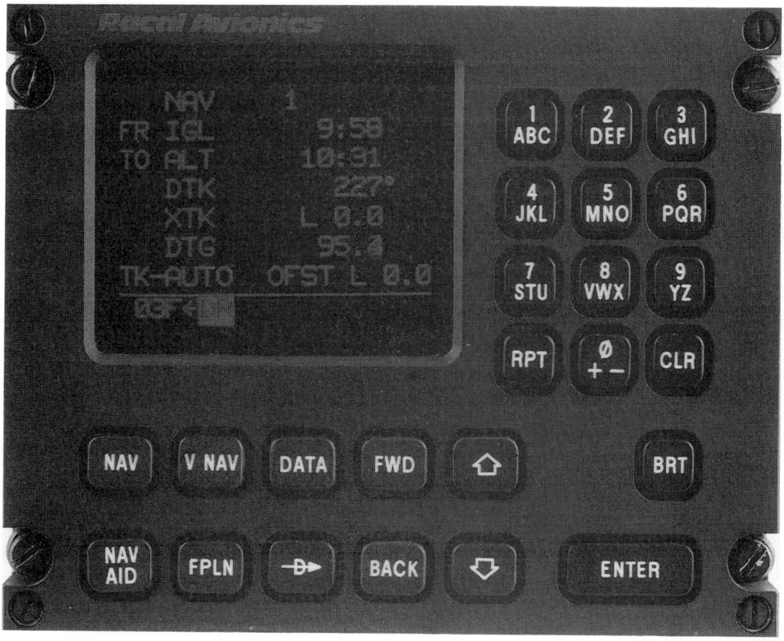

Plate 36 The R-Nav2 control display unit used with the Decca Mk 32 navigator equipment. *(Photo: Racal Avionics)*

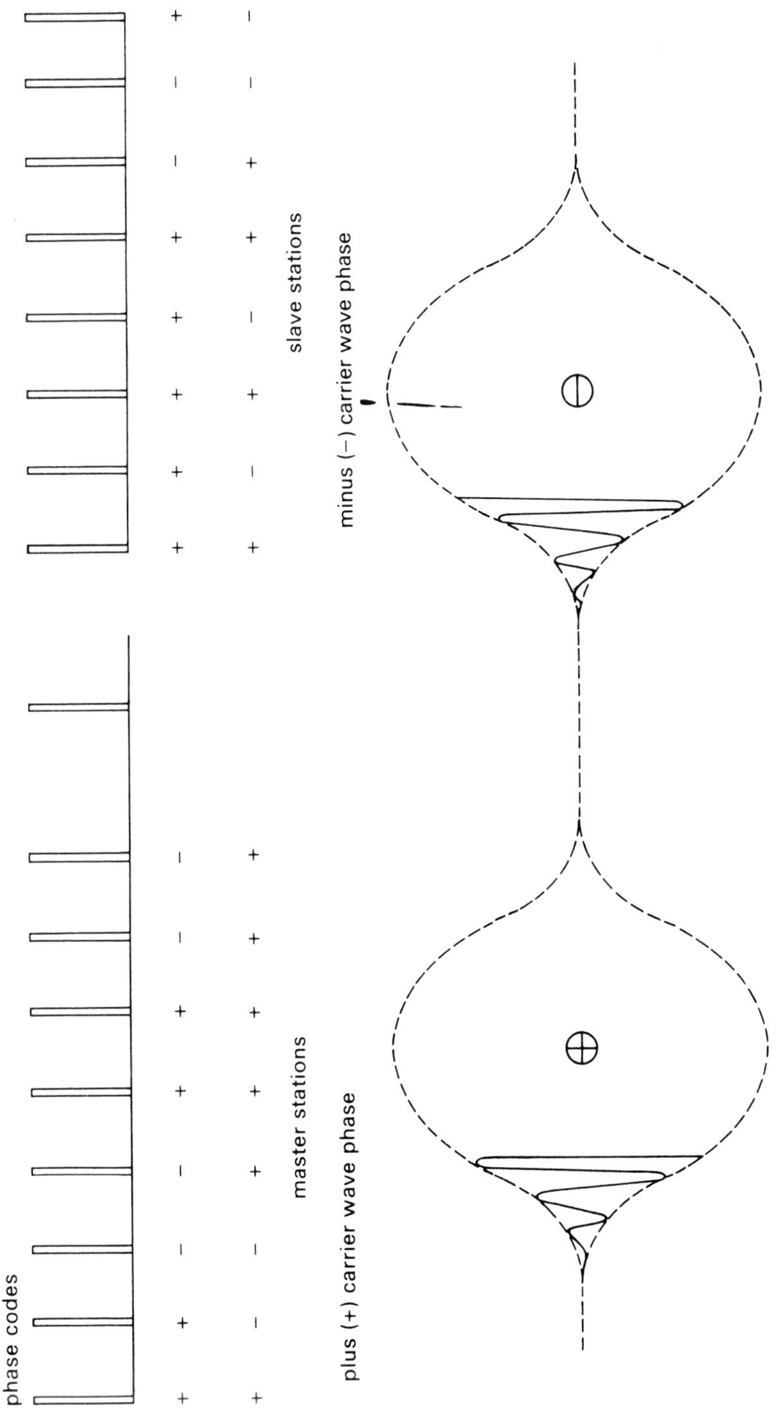

Fig. 48 Loran-C. Discrimination of master and slave stations.

power of four megawatts which is fed to a vertical mast radiator typically 1350 ft in height. A pulse transmission of such power could reasonably be expected to cause considerable interference to signals on adjacent channels. It is therefore specified that these transmitters confine 99% of their radiated energy within the spectrum 90 kHz to 110 kHz. This is achieved by arranging a slow build-up and decay of the transmitted pulse. On reception, sky wave contamination can be expected about thirty microseconds after the leading edge of the pulse is received. For this reason, more accurate equipments use only the first three cycles of the pulse at which point it has reached about 50% of its peak power.

The master station radiates a group of nine pulses, at a repetition rate of ten to twenty-five groups per second, the pulses within the group being spaced by 1000 microseconds. After a delay, which is in excess of the one-way propagation time between master and slave, plus a further 2000 microseconds to allow the sky wave to die down, the slave radiates an eight pulse group also at 1000 microsecond spacing. Still later, the next slave transmits a similar group.

Groups of pulses are transmitted as this method effectively increases the mean power output of the equipment without necessitating excessive transmitter powers. Each pulse within a group may be either in or 180° out of phase with an established reference, this phase coding enabling individual chains to be identified.

At the operational frequency of this system, radio frequency phase can be measured to an accuracy of about 0.03 microseconds. In consequence the accuracy of a fix is determined mainly by the geometric angle of cut of the position lines and the stability of propagation of the transmitted signal. In the past, propagation variations up to about 0.5 microseconds have been observed, however, these occur relatively slowly and can be partially compensated by slightly adjusting the timing between master and slave.

The Loran-C receiver

With all transmitters operating on the same frequency it may be thought that receiver design could be greatly simplified, however, the other requirements which must be met are severe and more than compensate for the relative simplicity of the tuning arrangements.

Within the 20 kHz bandwidth of the receiver, the required signal may often be as low as 20 dB below the noise. Other interfering signals, either pulse or cw may be 35 dB higher, furthermore, the signal strengths of the desired stations may vary by as much as 120 dB. Any receiver designed to operate under such hostile conditions must possess extremely high effective selectivity. Such selectivity is not obtainable with conventional passive filter circuitry, instead recourse is made to slow response servo loops with long integration times which track the received signals. Additionally automatic notch filters may be incorporated which act as both search and rejection elements. These continuously scan the receiver bandwidth to ensure that all new interfering signals are examined so that the rejection elements may be directed to the strongest interference. Modern equipment will acquire a signal at a level of 6 dB below atmospheric noise and maintain tracking down to a level of 20 dB below noise. Under such circumstances signal acquisition may take as long as 250 seconds but once locked on to the incoming signals the time differences may be indicated to an accuracy of one-tenth of a micro-second on a digital electronic display.

In common with other LF and VLF systems the aerial requirements for airborne Loran receivers is not excessive, the only parameter being that it be sufficiently large to ensure that the received atmospheric noise substantially exceeds that generated internally in the receiver. This can usually be met by aerials with an effective height of considerably less than a metre.

Loran-D

This system is intended for short range working at low altitudes. It is compatible with Loran-C and most airborne receivers will operate equally well with either system.

In meeting the differing requirement compared with Loran-C, the Loran-D system uses shorter baselines, and requires lower radiated power. This gives the advantage that smaller transmitters may be used in conjunction with lower (300 ft) aerial masts. In order to compensate to some degree for this, however, each transmitted group consists of sixteen pulses, 500 micro-seconds apart which are sampled at their peaks, instead of on their leading edge.

6.3 The Omega system

The Omega navigational system is being established by the USA Navy as a world-wide aid for aircraft, ships and submarines. Eight beacons are necessary for complete coverage, seven being operational at the date of writing and the eighth being planned for installation in south eastern Australia. The system is a hyperbolic aid operating in the Very Low Frequency (VLF) spectrum on frequencies between 10 kHz and 14 kHz (30 km to 20 km wavelength).

Propagation at very low frequencies is completely different from that at high frequencies, being a waveguide mode between the reflecting ionosphere and the earth below. The optimum mode of propagation at Omega frequencies is known as TM in which the electric field forms a half loop between earth and ionosphere. The extremely long range of the Omega signals is due to very low attenuation of this mode of propagation.

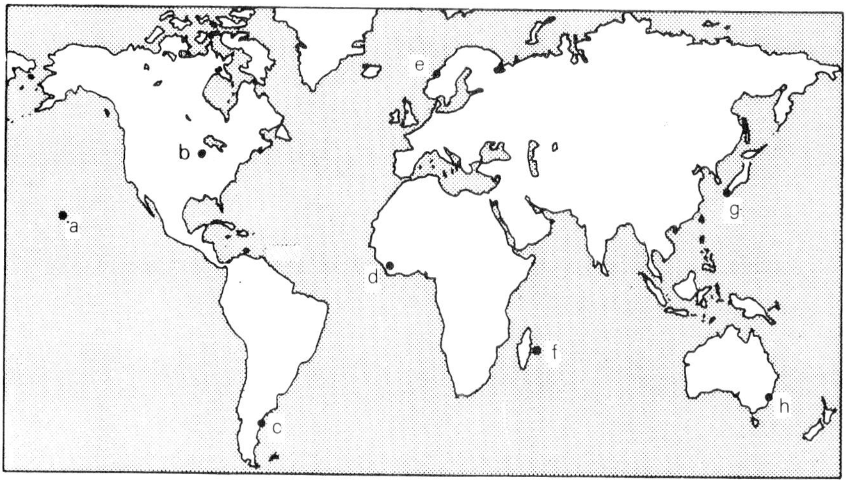

Fig. 49 Location of OMEGA stations, a–Hawaii, b–North Dakota, c–Argentina, d–Liberia, e–Norway, f–La Reunion, g–Japan, h–Australia.

The range and attenuation of Omega signals also depend on several other factors, including the path which is traversed by the signal and the angle which the transmission makes with the earth's magnetic field. Sea water paths offer least attenuation and ice paths the most. Signals over the Arctic and Antarctic are therefore limited in range. In daytime the ionospheric reflection height is about 50 km but at night this increases to 70 km or more resulting in lower attenuation with consequent increased range. The ultimate range of the transmission depends upon the ratio of signal to noise at the receiver and at these frequencies the noise is predominantly generated by lightning discharges in thunderstorms. The frequency and intensity of such conditions varies with latitude, season, weather and time of day thus range is also dependent on these factors.

Even where signals are strong, they may not be suitable for navigational purposes due to multimodal interference. The reason for this is that the transmitting aerial, although extremely large in physical terms, is, neverthless small compared with the operational wavelength. This causes many spurious modes to be excited within the vicinity of the transmitter. Most of these fade rapidly, but some, such as TM_2 (one and a half loops between earth and ionosphere), propagate to significant distances, interfering with the desired TM mode. Consequently an Omega signal cannot be used within 1000 km of the radiating station in daylight and for a proportionately greater distance at night.

Other regions where transmissions become unusable exist near the antipodes of the transmitter where direct and long path signals interfere and also where paths have traversed transequatorial regions towards the west.

Although not of concern to aviation it is interesting to note that VLF signals such as those used by the Omega system have the property of penetrating sea water to a considerable depth, the attenuation of one metre of sea water (4 dB) approximating to that of 1000 km of atmospheric path. Since underwater, both signal and atmospheric noise are attenuated equally, the signal to noise ratio will remain the same as on the surface, thus underwater range will depend on the sensitivity of the receiver. These frequencies can therefore be used to provide world-wide communications and navigational information to submerged submarines.

The Omega transmitter

The signal format of an Omega station consists of four bursts of signal, the first radiated on 10.2 kHz, the second on 13.6 kHz, the third on 11.33 kHz and the fourth on 11.05 kHz. The length of these bursts varies between 0.9 and 1.2 seconds dependent on the station and frequency in use, this being one of several ways by which emissions of the various stations may be identified.

It is also arranged that at each half minute, all carrier frequency currents in all transmitting stations pass through zero with a positive slope. To achieve this degree of synchronisation, four caesium frequency standards are located at each Omega station. Synchronisation update is accomplished weekly by computing each station's offset and divergence from an arithmetic mean. This mean is also compared to Coordinated Universal Time (UTC) based on the United States Naval Observatory master clock and an additional system correction is inserted such that 'Omega time' runs parallel to UTC. Using this system synchronisation can be achieved to an accuracy of two to three microseconds. Experimental work is continuing to provide a basis for more accurate Omega system synchronisation techniques.

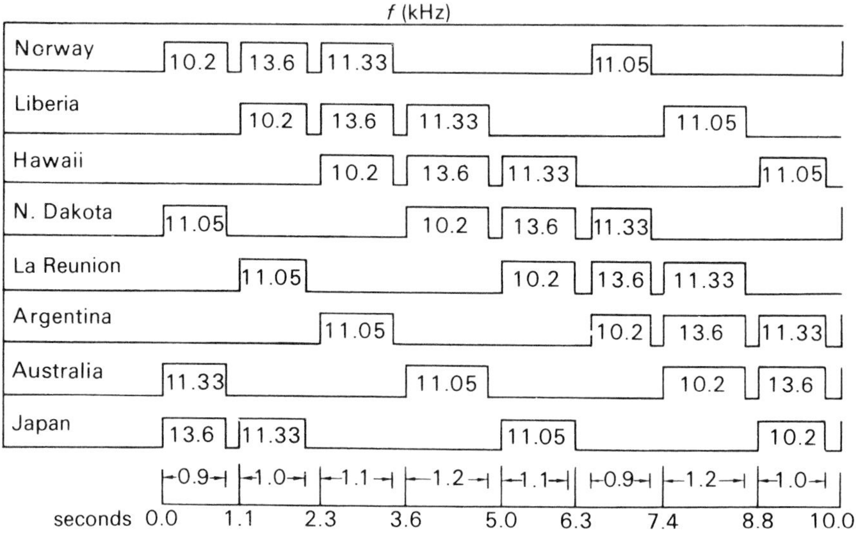

Fig. 50 Omega signal transmission format

The final amplifiers of the transmitter give an output of 150 kW but as due to physical limitations, the aerials being necessarily electrically small, the actual power radiated is in the order of 7% of this figure. Two types of aerial are used, valley-span systems being in use at the stations at La Reunion, Hawaii and Norway and 400 m vertical towers at North Dakota, Argentina, Liberia and Japan.

The receiver

Receiver designs are continually evolving, nevertheless the basic principles of operation of all receivers are the same: the phase differences between the signals from different stations measured, the propagation corrections applied and from these the position is deduced. With cheaper equipments the later steps are done manually whereas the more expensive equipments are fully automated.

At very low frequencies, noise levels are high and thus to receive signals at long range, receivers must be extremely selective. The useful range of Omega signals has been defined as the distance at which noise and signal voltages are the same in a 1 Hz bandwidth receiver but of course the range for any particular receiver depends on its quality. Signal and atmospheric noise levels are normally strong giving, for example, a signal strength of thirty microvolts per metre and a noise voltage of three microvolts per metre at a range of 8000 miles from a transmitting station into a receiver of 10 Hz bandwidth.

Operation

When three or more Omega stations can be received, normal hyperbolic techniques may be used for position fixing. Using the 10.2 kHz transmissions similar signals are received on lines of position spaced at halfwave (i.e. 15 km) intervals but unless the navigator is also dead reckoning or lane counting his position can remain undetermined, especially if he has been out of contact with one of the stations for a period of time. The use of a multiple frequency receiver can reduce these ambiguities for the lanes derived from the 13.6 kHz signal only coincide with those of the 10.2 kHz signal every third lane, reducing the ambiguity to 45 km. The subsequent use of the 11.33 kHz signal extends the ambiguities three times further to 135 km. The effective time of arrival can provide a further aid, as can the continuous tracking of at least four stations. Receivers may also be designed to operate on comparison frequency of 3.4 kHz being the difference between the 13.6 kHz and 10.2 kHz transmissions and which gives a 45 km lane width.

Where only two can be received, normal hyperbolic techniques are not possible but a method known as range-range may be applied. This requires a receiver incorporating a precision clock and given intersecting circles of position, instead of hyperbolae, and increases the ambiguity lane width to one wavelength (30 km).

A review of the error mechanisms in the Omega system indicate that the most significant errors result from variations from the so-called normal ionospheric conditions and from anomalies caused by solar flares and the geomagnetic related events. Two techniques have been proposed which may overcome many of these effects: composite Omega and differential Omega.

In composite Omega, mathematical analysis has indicated that if weighting factors are used to combine the 10.2 kHz and 13.6 kHz transmissions in order to obtain a 3.4 kHz signal, more accurate results should be obtained than by use of the 3.4 kHz difference frequency.

In differential Omega, a receiver is installed at a known ground site. The difference between the received phase reading at the site and the phase reading calculated for that site are transmitted to the aircraft or other user. By this means, errors due to ionospheric or anomalous propagation effects are greatly reduced.

Section 7
Airborne systems

7.1 Inertial navigation

Before the advent of radio aids to navigation, the position of a ship or aircraft was determined by maintaining a record of track and estimated velocity, plotting these parameters on a chart and thus deriving the contemporary position. The errors introduced by inaccurate determination of velocity, wind or ocean currents were corrected by either celestial observation or at the first landfall.

With the introduction of radio techniques, direction finding and position fixing systems such as Consol, Loran and more recently Omega came to the aid of the navigator. Such radio systems, however, are limited by the fact that they require extensive networks of ground stations and are subject to both natural and man-made interference.

In recent years, developments in both accelerometers and gyroscopes have enabled a return to the earlier method of dead reckoning with an automatic system which is capable of an accuracy comparable with radio derived systems. This system is called the inertial navigator since it makes use of the laws of motion first postulated by Sir Isaac Newton.

Provided that the initial position and alignment of the vehicle is entered into the inertial navigation equipment, any movement of the vehicle will be sensed by accelerometers attached to a gyroscopically controlled reference table. The associated computer accepts the signals from the accelerometers, compares them with a timing signal and makes a mathematical calculation to determine the current position of the vehicle. This may be displayed to the pilot in any convenient format.

Additionally, further computing power enables a predetermined journey to be programmed into the computer, the output of which will control the autopilot to cause the aircraft to fly the predetermined tracks. Arrangements may also be made for the navigation computer to accept alternative inputs from other navigational aids to cross-check the inertially derived positional data.

The basic principles of inertial navigation

The basis of inertial navigation lies in the three laws of motion originally postulated by Sir Isaac Newton nearly three hundred years ago. These are:

(a) A body continues in a state of rest, or uniform motion in a straight line, unless it is acted upon by an external force.

(b) The acceleration–rate of change of velocity–of a body is directly proportional to the force acting on the body and is inversely proportional to the mass of the body.

(c) To every action there is an equal and opposite reaction.

While totally reasonable to Newton and his contempories, his use of the term 'at rest' eventually came under heavy criticism for when Einstein published his paper on the special theory of relativity in 1905, he totally destroyed the premise of 'absolute motion'. The substance of new theory was that nothing is at rest and that the term 'at rest' meant merely that the object under observation was moving at the same velocity as some other object, its coordinate system and the observer.

The primary measuring device in an inertial navigation system, the accelerometer, demonstrates this theory for it makes no distinction between 'at rest' and any other fixed velocity. It does, however make distinction between truly fixed velocities and those which we may regard as fixed, but are really fixed speeds along curved paths.

Velocity is a description of a state of motion, being composed of both speed and direction, therefore if the speed of an object is constant but its direction is changing, so then is its velocity. A change in velocity, be it in either magnitude (speed) or the direction of motion, is an acceleration (or deceleration). A body accelerates (or decelerates), i.e. changes its state of motion, only if it is acted upon by an external force.

All matter tends to retain its existing state of motion and consequently resists any changes to that state. This property of matter is known as inertia. The amount of inertial force which will be displayed by a body is proportional to the amount of force being exerted upon it and the acceleration rate which the body will suffer in response to an external force is proportional to the magnitude of that external force. It therefore follows that the rate of acceleration of a body is proportional to the magnitude of the inertia. If the inertial force can be measured, then the rate of acceleration is known.

The inertial force displayed by a body under a given rate of acceleration gives a measure of that body known as mass. Mass is equal to the force divided by the acceleration.

In developing an inertial navigation system it is possible, by application of Newton's second law, to construct a device which is capable of detecting minute changes in velocity. This is the accelerometer.

Accelerometers

As implied by their name, accelerometers are devices capable of measuring acceleration. Although there are today many varieties of these instruments, all work on the same basic principle, that is, the measurement of inertial 'pushing back' of a known mass when acted upon by externally applied forces.

In its most simple form the accelerometer may consist of a case in which a mass is suspended in such a way as to permit some degree of restricted

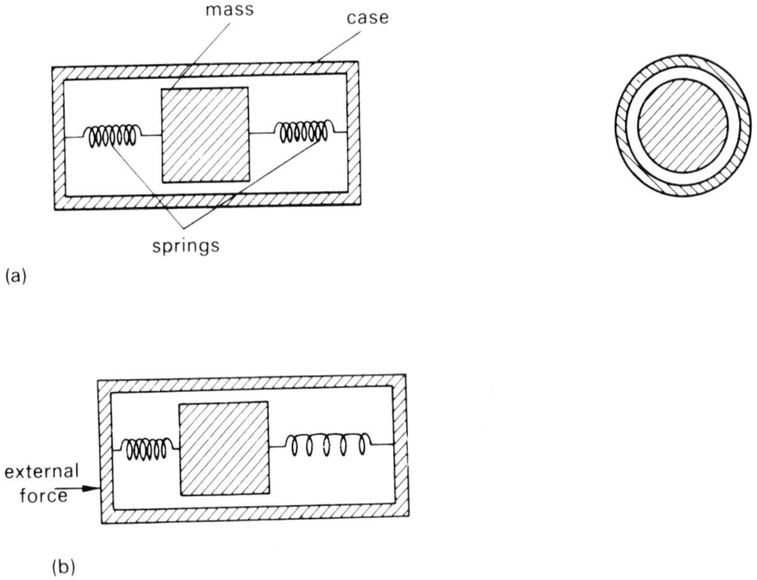

Fig. 51 (a) Simple accelerometer at rest. (b) Under acceleration.

movement between the two in a single direction. Consider first, therefore, an example where the case is a section of tubing and the mass is an internal piston. The piston is restrained in a central position by two springs located along the axis of the tube. The piston will therefore be free to move along the axis of the tube but in doing so will be restrained by the action of the springs.

When in a state of rest, the position of the mass within this simple accelerometer will be dictated by the restraining springs alone. If a force is applied longitudinally to the case of the instrument, this will move but due to inertia, the mass will attempt to remain stationary. The restraining springs will apply a force to the mass and eventually it will be forced to move, but due to the forces on the springs, its movement will lag behind that of the case. With constant force applied to the case, acceleration will be constant but at some point in the acceleration, the forces exerted by the springs on the mass will equal this force. A point of equilibrium will then be reached at which a constant displacement will exist between case and mass, this being proportional to the acceleration of the mass, its case and any object to which they are attached.

An alternative form of accelerometer uses a pendulum, in which case the displacement increments are angular rather than linear.

The relative movement of mass and case in most accelerometers is very small–so small, in fact that the displacement can only be detected by electrical pick-offs. These usually incorporate a pair of primary coils mounted on the case and a secondary coil attached to the mass. The two primary coils are spaced either side of the null position, each being fed with a signal of the same frequency but in opposite phase. The secondary coil is fixed to the mass itself in such a position that at the null it receives equal and opposite signals from both primary coils which results in no output. When subjected to acceleration the mass moves into the area dominated by one or the other primary coils, thus developing an output signal, the amplitude of which is proportional to the movement and the phase indicates the direction.

In a typical application the output of the accelerometer is amplified and fed to a phase sensitive demodulator, the output of which is a d.c. voltage of a level and polarity indicating the degree and direction of the acceleration.

In the basic accelerometer so far described the use of springs introduces a non-linearity problem. If the acceleration is increased in equal increments the mass's displacement will build up in diminishing increments until the expansion–contraction limits of the springs are reached. This limitation has been overcome by a technique known as force-rebalancing.

In using the force-rebalancing principle, electromagnets are fitted to the case of the accelerometer, positioned to react with permanent magnets fitted to the mass in such a way as to return the mass to the null (zero acceleration) point. The demodulated and amplified accelerometer output signal is fed to the electromagnets, known as torquers, and afterwards dropped across a resistance to develop an analogue voltage of acceleration. Therefore, instead of measuring displacement, this method measures the current required to develop a force capable of resisting displacement. The end result is the same–a signal proportional to acceleration.

In most such systems, an actual proportional displacement exists, but in others the d.c. amplifier feedback loop is arranged to give an increasing output until the mass has been returned to the null point.

There are, of course, many other factors in accelerometer design. These are mainly concerned with overcoming the mechanical limitations of the system and as such do not effect the basic requirement which is to produce an output signal proportional to the acceleration of the body to which the accelerometer is attached.

Timing

Having measured the acceleration of the aircraft, in order to determine the final velocity and subsequently the distance travelled, it is necessary to measure the duration of that acceleration. This is effected by the incorporation of an internal electronic clock the accuracy of which is determined by an oscillating high frequency quartz crystal in a similar fashion to the quartz controlled watches and clocks which have recently found favour in the consumer market.

Displacement calculation

If the accelerometer produces a true indication of acceleration to the computer, then the application of calculus in the form of a double integration with respect to time will produce accurate information concerning the movement of the accelerometer along its sensitive axis. If the accelerometer is fixed to the frame of the vehicle and aligned in the direction of movement, the computer output will indicate the distance travelled.

If two accelerometers are with their axes at right angles, any change in motion in the plane of their axes will be sensed by one or other or both, each one reacting to one of the two components of movement. When such an accelerometer system is made to maintain some known angular relationship to the terrestrial coordinate, i.e. latitude and longitude, the accelerations sensed and interpreted in the reference system can be related to the other. The simplest relationship is where one accelerometer is aligned north and south and the other east and west. In such circumstances the acceleration, velocity and distance components are directly translatable to the components in the terrestrial system. Any other relationship is also useable. Provided that the precise relationship is known at all times, the navigation system computer has the capability of calculating the necessary coordinate transformation.

To enable such a system to be used for navigational purposes, a specific relationship of the highest degree must be maintained between the reference and terrestrial coordinate axes.

Although the accelerometer assembly is suspended in a moving vehicle, it must remain angularly fixed or rotated at a very precise angular rate. To achieve the necessary stability it is necessary to make use of the gyroscope.

Gyroscopes

The basic principles of gyroscopes have been studied since the mid-eighteenth century. The mathematician Eular first examined the behaviour of spinning rotors about the year 1750 and a hundred years later the French scientist Foucault used them in experiments demonstrating the rotation of the earth. Foucault named his device the gyroscope from the Greek words *gyro*, to turn, and *skopein*, to view or see.

The first commercial applications for navigation came when the gyrocompass was patented by a Dr Kaempfe in Germany in 1908 and by Dr Sperry in the USA in 1911.

A gyro may be described as a spinning mass, usually a wheel or disc, turning about an axis, supported by a gimbal system which allows the mass to remain in the same alignment no matter what angular position is described by the gimbal system support.

The wheel is constructed of either stainless steel, titanium or beryllium, the latter being more favoured in recent years due to its mechanical stability. In designing a gyro wheel the requirement is to provide the greatest possible moment of inertia coupled with adequate mechanical strength and heat dissipation. To achieve the necessary angular momentum, the wheel is driven by a gyro motor, usually a synchronous, hysteresis type which is capable of providing a precision drive when energised by a power supply of high frequency stability.

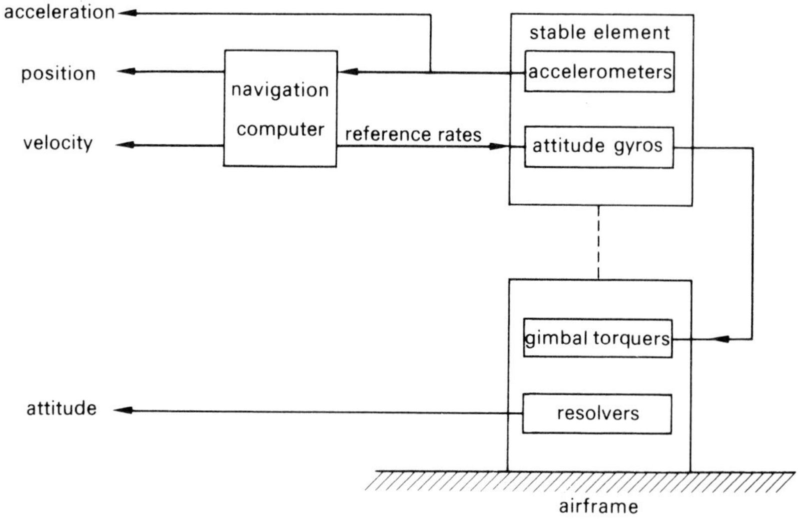

Fig. 52 Conventional INS – schematic

The rotating gyro wheel is mounted within a gimbal system which is arranged to permit the wheel to maintain a constant angular position regardless of the attitude of the platform to which the gimbal system is attached. To make use of the stability of the gyro wheel, it is necessary to transfer this stability to some form of stable platform to which the accelerometers described earlier may be attached. This is achieved by fitting the gimbal mount to a platform and transducers to register movement of the gyro inner gimbal. The output of the transducers is fed to a servo amplifier and motor which maintains the platform which is itself mounted on gimbals in a position controlled by the gyro. By fitting the platform with two gyros mounted at right angles, its position is controlled and the requirement for a platform stable in all axes is met. To this can be fitted the accelerometers, the output from which is fed to the navigation computer for computation of the aircraft position.

Types of inertial navigation systems

The analytic inertial navigation system

This system uses a platform with a fixed angular reference to some point in inertial space. Because the platform remains rigid in space and rotates around the earth, the output accelerations become extremely complex, however, these may be resolved into two major accelerations; that due to vehicle movement and that due to gravity. For navigational purposes only the former is required and the latter must be cancelled out. Due to the fact that the earth's gravitational acceleration varies widely, this requires that a vast amount of data be stored within the computer to effect cancellation.

Within the navigation computer, the outputs of the accelerometers are summed and acceleration corrections applied to produce the true acceleration of the vehicle. The result thus obtained is then integrated twice to derive displacement. From this information a coordinate converter calculates the aircraft position in latitude and longitude which is then displayed in a convenient manner for the pilot.

The semi-analytic inertial navigation system

This is the most common system in use today for the main advantage over other systems is economic. The platform gimbal system is simple and the computer may be either analogue or digital.

In this system the platform is aligned normal to the gravity vector and is not necessarily aligned to true north. The north accelerometer output is corrected and integrated with respect to time to derive the north velocity component of the vehicle's track. Through scaling this is converted to an angular velocity and integrated again to give a position read-out in latitude. In similar fashion the east component of the vehicle track is derived and processed to obtain a longitude read-out of position.

Strap down inertial navigation systems

The Strap Down inertial system does not use a gimbal mounted reference table. The gyros and accelerometers are mounted directly on the vehicle frame and the reference table is, in effect, replaced by the computer. The system was first described by W. Newell in a patent in 1956, but development was delayed because analogue computers are not sufficiently accurate and at that time digital computers could not be constructed in a suitably compact form.

In this system, the gyros supply angular rate signals to the B matrix which develops directional cosine signals used to indicate the aircraft attitude with reference to an inertial reference frame. A coordinate converter accepts inputs from the accelerometers and the B matrix derives accelerations along the inertial reference axes. These signals are fed to a position computer for derivation of cartesian coordinates representing the aircraft's position in inertial space which are then summed to provide latitude and longitude read-outs by a vector solver.

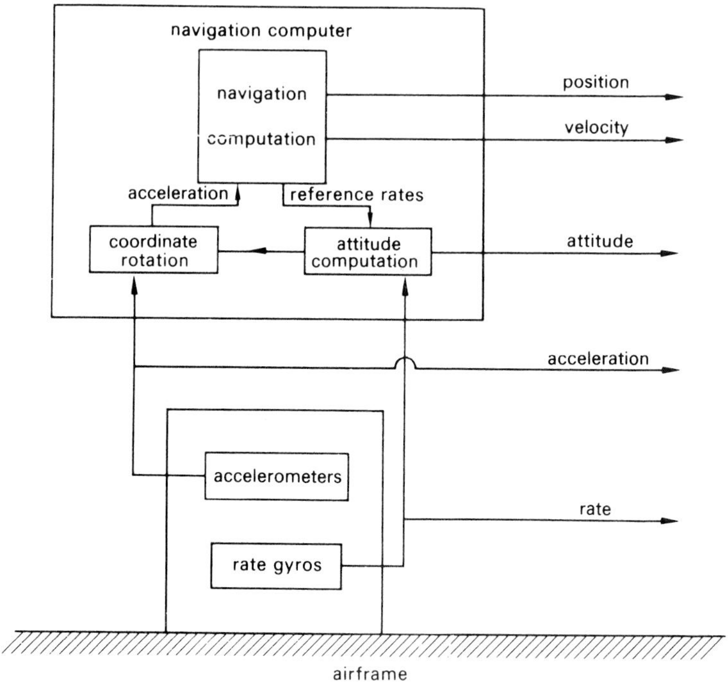

Fig. 53 Strap down INS – schematic

The accuracy of the Strap Down inertial system is limited by that of the inertial sensors and the computer incorporated within the system.

Hybrid inertial navigation systems

All navigation systems are subject to errors, and in this respect inertial navigation is no different from any other. By combining one or more navigation systems it is possible to obtain an overall navigational accuracy greater than that possible from any of the component systems.

Typically the positional information derived from an inertial system may be compared with that from ground based aids, Doppler radar or celestial navigation techniques. Two methods of updating are used. The first compares inertial velocities with those derived from some other system and uses the error voltage between the two to damp out platform errors. The second, known as the reset method, uses information from another navigational system to reset the position of the velocity shafts, ignoring the position of the platform.

Operation

The output signals from the accelerometers of any inertial navigation system, when processed by the computer and compared with the internal clock, are capable of indicating the displacement of an aircraft from the point of departure with a high degree of accuracy. However, to use this information for practical navigation, it is necessary to relate this displacement to the geographical coordinates of the point of origin of the flight. It is therefore a vital ingredient of the pre-flight procedure to insert the geographical coordinates of the aircraft in latitude and longitude into the INS equipment. Once this information has been input, it is essential that the equipment remains switched on, for should the system be switched off, reference to the point of alignment is lost and re-alignment is not possible in the air. It is not necessary to insert any directional information into the equipment for the accelerometers to sense the rotation of the earth and, from this, stabilise the platform with respect to true north from whence all other angular measurements are derived. For this reason, during the alignment procedures the aircraft must remain stationary, although minor vibrations, due to passenger and cargo loading or light gusty winds will not affect the procedure.

In addition to the coordinates of the point of alignment, those of a number of other locations where course changes will be made may also be input. These are known as waypoints. When in flight, if the INS is coupled with the autopilot, the aircraft will automatically fly from one waypoint to the next on a great circle route, accomplishing the necessary course changes for the consecutive track legs. When required, the INS will also provide the necessary guidance for flying a course parallel to that set in the flight plan. For further assistance to the pilot, the INS will also display other navigational information such as: track angle and ground speed; heading and drift angle; cross track distance and track angle error relative to the desired track; present position; distance and flying time to any or between any two waypoints; and the total distance and flying time remaining in the programmed flight plan.

Ring laser gyros

Another recent development in this field is the ring laser gyroscope.

This device has several characteristics which make it suitable for use in inertial navigation, particularly with Strap Down systems. These may be summarised briefly as:

no moving parts
no warm-up period required
high long-term stability
high rate capability
clearly defined input axis
insensitive to acceleration
fine resolution over full dynamic range.

Operation

In the understanding the operation of a ring laser gyro it is first necessary to consider the situation when an electromagnetic radiation is being emitted from a moving source. As the velocity of such radiation is an absolute constant, the effect of source movement is that the static observer notices an apparent frequency shift due to Doppler effect. When the source of radiation is approaching, the observed frequency is higher and when receding lower.

Now consider the case of a beam of coherent light which is reflected around a triangular track by three mirrors mounted on a table. Should the table rotate in the same direction as the beam, the frequency will increase and the converse will also be true. Thus if the frequency change of the light beam due to rotation of the table can be measured, this will provide a measure of the rotation of the table. This is achieved by passing a second beam of coherent light, emanting from the same source, around the track in the opposite direction to the first. Whenever the table rotates, therefore, the beam traversing the path in one direction will increase in frequency and the other will decrease. The frequency difference is measured by allowing a small percentage of the coherent light to pass through one of the corner mirrors. A prism is used to reflect one of the beams such that it crosses the other in almost the same direction at a small angle (wedge angle). Due to the finite width of the beams, the effect of the wedge angle is to generate an optical fringe pattern in the read-out zone.

When the frequencies of the two beams are equal, i.e. under zero rate conditions, the fringes are stationary. On rotation of the table, the difference in frequency between the two beams causes the fringe pattern to move at a rate and direction proportional to the frequency difference, i.e. proportional to the angular rate.

Moreover, the passage of each fringe indicates that the integrated frequency difference (integrated input rate) has changed by a specified increment. Consequently, each fringe passage is a direct indication of an

Plate 37 A Litton Aero Products ring laser gyro block. Unlike most other manufacturers, these use a four-sided light path.

Plate 38 The Litton Aero Products laser gyro equipment. Three gyros are included, one in each plane.

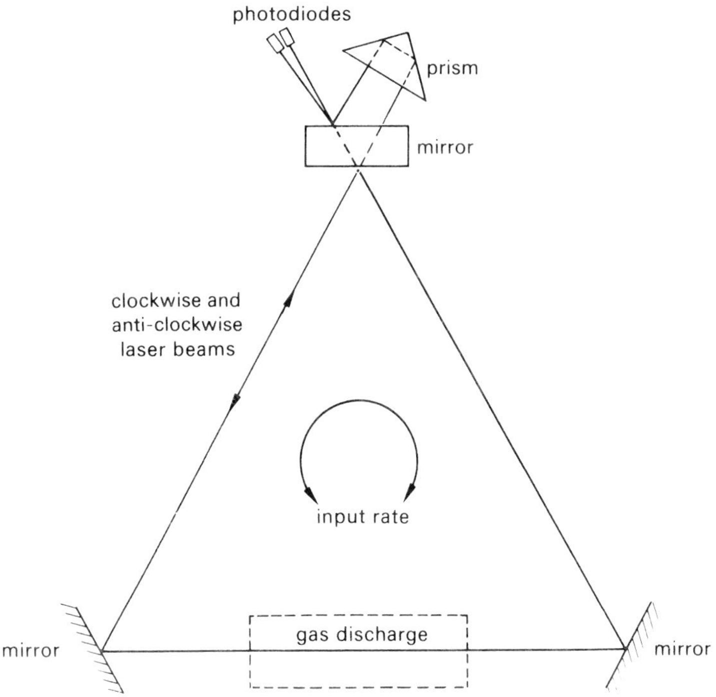

Fig. 54 Principle of the laser gyro

incremental integrated rate movement–the precise form of signal required for a rate-gyro Strap Down navigational system

To detect the fringe movement, two photodiodes are mounted 90° apart (in fringe space) in the fringe area. As the fringes pass by the diodes, sinusoidal signals are generated whose frequency and relative phase indicate the speed and direction of rotation. The sinusoidal outputs are then converted into digital form for the computer input by simple digital pulse-triggering and direction-logic circuits.

To achieve the necessary degree of path length stability the optical cavity is constructed of a material of extremely low coefficient of expansion. A single structure contains the helium-neon gas with the lasing mirrors and electrodes forming the seals. A high voltage applied across the electrodes ionises the helium-neon gas mixture, thereby facilitating the laser action to provide the coherent light.

Although simple in concept, nevertheless there are several basic sources of error in a laser gyro, the most fundamental of which is possibly frequency synchronisation or lock-in. To achieve the necessary light path, the beams have to be reflected at least three times. As no mirror is perfect, a small amount of energy is back scattered at each reflection in the direction of the other light beam. This is the main contribution to a coupling effect between the two light paths which tends to cause the waves in opposite directions to lock together when gyro rotation is below a threshold value. This is at a rotation of about 6° per minute and formed a major problem in the design of navigational quality gyros.

Two alternative systems are used to overcome the lock-in effect, each of which biases the gyro out of the dead band by mechanical movement.

In the first of these, the gyro is rotated to and fro through 360° at a predetermined rate, whilst in the second, the gyro is vibrated at 400 Hz by means of a quartz transducer mounted at the centre of the gyro block. In the latter case the movement is barely perceptible to the eye, but nevertheless it represents a considerable movement in terms of the wavelength of the laser.

In each case, compensation for the rotational bias is included in the computer software.

Experience in recent years has proved the laser gyro to be both accurate and reliable. It is therefore conceivable that within the foreseeable future, mechanical gyros will be totally superseded by laser equipment in aircraft navigational systems.

Future developments

The electrostatic suspended gyro

One of the main problems in the development of conventional gyros is that of bearing design. However, a device which would appear to overcome this problem, the electrostatic suspended gyro, is currently under development. This consists of a ball which is spun within a closely surrounding spherical case without there being any mechanical connection between the two. The spinning ball is therefore free to reflect the total angular motion of the vehicle to which the device is attached.

The rotor consists, typically, of a thin shell of beryllium with an internal flange to provide a preferred axis of rotation. This is electrostatically supported within the casing usually by three pairs of electrodes, which, by means of a ball-position sensing device and a servo system, maintain the ball centred by exerting attractive forces. Frictional losses are reduced to a minimum by evacuating the casing space to a high degree of vacuum.

By use of such techniques, and because the electrostatic forces are sensibly normal to the surface of the ball, drag which would tend to slow the ball down is almost entirely eliminated and once the ball has been spun up, operational speeds can be maintained for several weeks without further drive.

The relative movement between ball and case due to aircraft movement is detected by the painting of a suitable pattern on the outer surface of the ball. This is sensed by a number of optical detectors whose output signals can be used to determine the full rotational motion of the ball.

The electrostatic gyro is a relatively new concept which, however, presents considerable manufacturing difficulties. Despite this, the device offers considerable potential for future use in Strap Down inertial navigation systems.

The fibre-optic gyro

Development is currently proceeding on the development of a fibre-optic gyro (FOG). Operating in a similar manner to the laser gyro, the FOG is a rotation rate sensor which uses optical fibre as the propagation medium for the light. This technique is relatively inexpensive, for it eliminates the extremely expensive block which is an essential part of the laser gyro.

One of the leading companies in this field is Standard Electrik Lorenz (SEL) who are currently evaluating the system for aircraft, land vehicles and tracking systems.

The SEL system consists of a source module which comprises a multimode laser diode, a monitor photodiode, a temperature sensor, a thermoelectric cooler and a laser diode to fibre coupling assembly. The system works in the 820 nm region.

The output from the source module passes via a 3 dB optic coupler using a fused taper technique to an integrated optical circuit which provides the functions of beam splitter, optical phase modulator and polariser. The components for this are based on titanium-indiffused single mode channel waveguides on a dielectric lithium niobate substrate. Reflection-free fibre-to-waveguide interfaces are used.

The light path comprises 100 m of single mode polarisation preserving fibre which are coiled into a carrier and encapsulated. The carrier is rigidly mounted in the housing in order that the reference plane may be defined.

The optical path is completed via the coupler to a detector module incorporating a PIN photodetector together with a hybrid preamplifier.

The optronics assemblies are completely contained within a hermetically sealed metal housing which is filled with inert gas.

7.2 Doppler navigation

The Doppler navigation system is a self-contained dead reckoning system which obtains the desired information through a measurement of aircraft velocity (both laterally and longitudinally) by means of Doppler radar and measurement of direction by means of a directional sensor such as a gyro or magnetic compass. The two sets of information are then processed in a computer in which the velocity is integrated into the two components of distance travelled from the point of departure.

Airborne Doppler sensors

Doppler sensors depend for their operation on the well known effect first documented by, and named after, the Austrian physicist Christian Doppler. This effect is noticed whenever a wave motion–sound or electromagnetic–is being observed and there is relative motion between the source and the

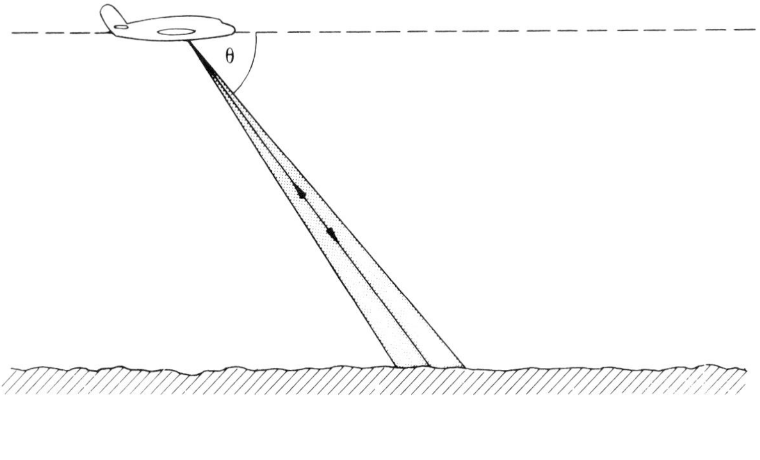

Fig. 55 Doppler forward beam depression angle

observer of the motion. Under such circumstances an increase in frequency will be observed when source and observer are converging and a decrease when receding. Furthermore, when the frequency of the source and the relative velocity of source and observer are known, the frequency shift (known as the Doppler shift) can be calculated. It therefore follows that if the frequency of the source is known and the Doppler shift can be measured, the relative velocity of source and observer may be deduced.

The utilization of this effect in the realisation of a practical velocity sensor for navigational purposes may be understood by first considering the situation where an aircraft transmits a signal in the form of a narrow beam of radio waves such that the radiation will hit the ground a short distance ahead.

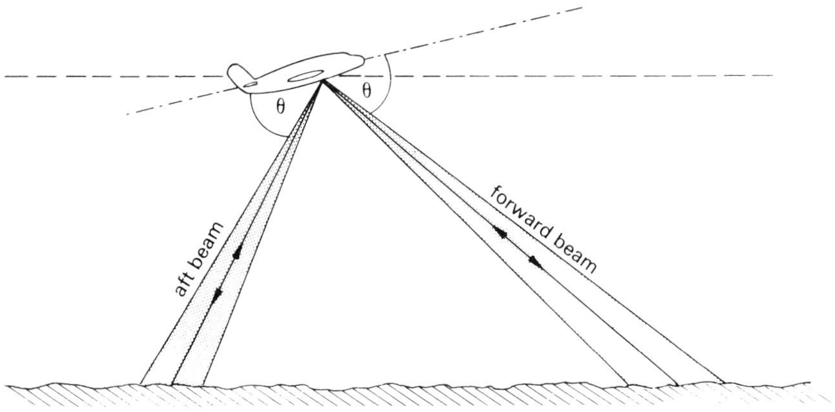

Fig. 56 Fore and aft beams to overcome pitch errors

On reaching the ground the energy will be scattered by the irregularities of the surface, some however, will be returned in the direction of the aircraft which will be picked up by the airborne receiver. On the path between aircraft and ground, the source will be approaching, thus causing the frequency of the signal hitting the ground to be higher than that radiated. The reflected signal will be received by an approaching observer, causing yet further frequency increase. On reception, the frequency shift of the reflected wave is measured and from this the relative velocity determined. To determine the true aircraft velocity, however, the angle of depression of the beam must be allowed for in the calculation. It is therefore essential that the angle be accurately known otherwise cumulative errors will appear in the integrated values of computed distance.

To ensure accuracy with the system so far described, the aircraft would have to be flown 'straight and level' at all times. As this is not possible, means have to be devised to compensate for any pitch changes. One such method is to mount the aerial on a gyroscopically stabilized platform such that it would maintain a constant attitude relative to the vertical irrespective of the attitude of the aircraft. A more economic alternative would be to transmit a second beam aft at the same angle of depression as the forward beam. By comparing the frequency shifts of forward and rearward beams, any errors due to pitch angle can be eliminated. The technique of using two such opposing beams is called the Janus configuration after the Roman god of doorways who was able to face in both directions at the same time.

It has been assumed so far that the heading and track of the aircraft are coincident. This, however, rarely occurs due to cross winds. Under such circumstances the velocity measurement derived from a single forward facing beam would be in error by a factor proportional to the angle of drift. A third beam is therefore introduced to provide a further velocity measurement.

The three beams are often arranged in the shape of the Greek letter lambda (λ), two beams being radiated forward symmetrically about the centre line of the aircraft and the third pointing aft, 180° displaced from one or other of the forward beams. Alternatively four beams may be radiated in an 'X' configuration. Although this fourth beam is not strictly necessary, it does give the advantage of redundancy which permits automatic monitoring.

The velocity measurement from each of these beams is fed to the computer, which, by allowing for the relative bearing of each of the beams relative to the axes of the aircraft, and comparing with a heading reference such as a gyro compass, performs a mathematical calculation deriving the velocities along track, of drift and of height variation. These results may be further integrated with respect to time to determine the displacement from the original point of departure. Thus, by ensuring that the point of departure is accurately referenced before take-off, the current position of the aircraft may be displayed in latitude and longitude. More complex computers have the facilities for inserting waypoints and displaying distance and bearing to next waypoint and updating referencing using ground derived aids.

Frequency and signal characteristics

The sensitivity of a Doppler radar (Hz per knot) increases with frequency. Furthermore, the higher the frequency, the narrower the beamwidth for a given aerial size.

If, however, the frequency is increased excessively, absorption, back scattering effects and rain reflections become an increasing problem. The preferred frequency is therefore a compromise and as a result Doppler

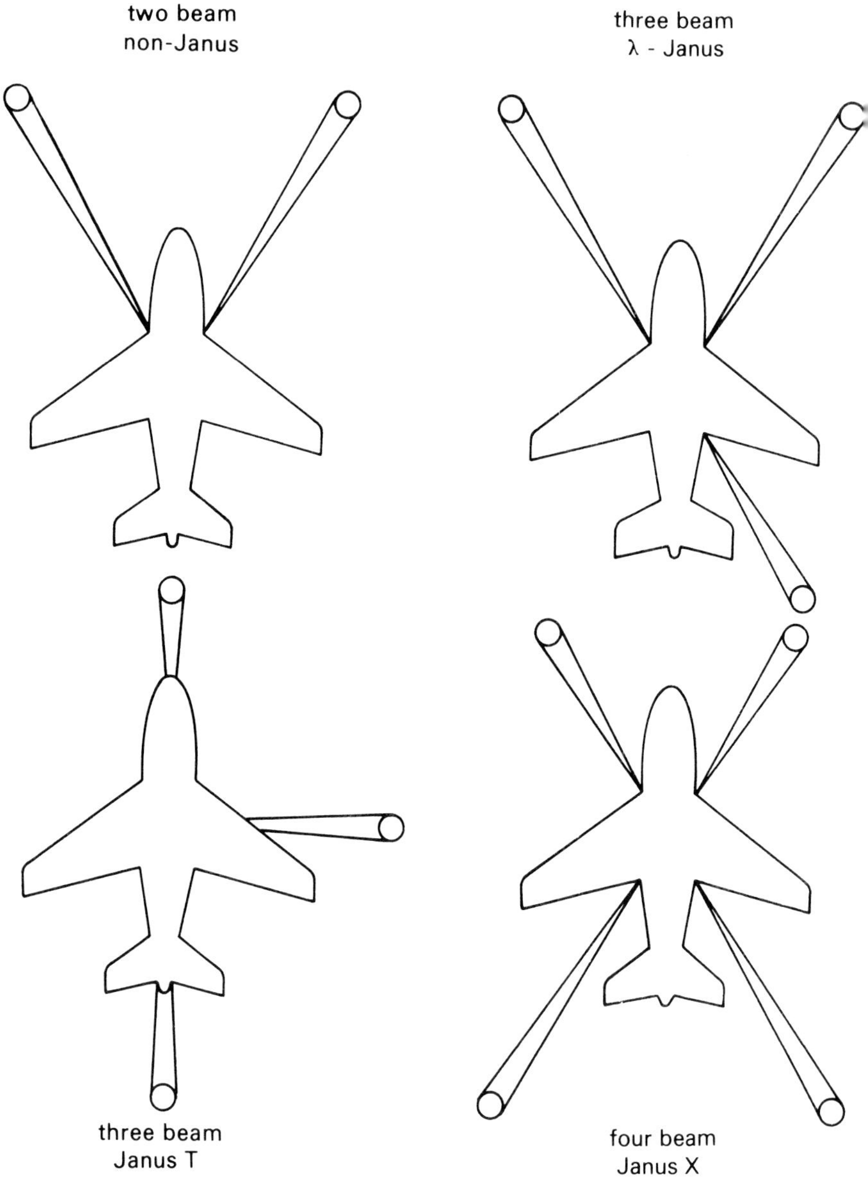

Fig. 57 Doppler radar beam configurations

airborne radar equipments operate either in X-band (8.8 GHz – 9.8 GHz) or Ke-band (13.25 GHz to 13.40 GHz).

Typical Doppler errors over land and after 10 nm of travel are of the order of less than 0.25% in ground speed and drift angle but any errors in the heading indicator will adversely affect the overall system accuracy.

The mathematics of Doppler radar

The basic calculations concerned with Doppler velocity sensors are not difficult and may be of interest to mathematically-inclined readers.

The frequency shift due to Doppler effect may be derived from the equation:

$$f = \frac{FV}{c} \quad \text{or} \quad f = \frac{V}{\lambda}$$

where f = Doppler shift
V = velocity of vehicle
F = frequency of transmission
λ = wavelength of transmission
c = propagation velocity of radio waves

Since the aircraft is not flying down the beam, the Doppler measured velocity V must be resolved into the aircraft's forward velocity V_h.

The basic expression now becomes:

$$f = 2\frac{FV_h}{c}\cos\theta$$

or $V_h = \frac{f\lambda}{2}\sec\theta$

where θ is the angle of depression of the beam.

The factor 2 appears because both transmitter and receiver are moving with respect to earth.

Fore and aft beams

If two beams are being rated in Janus configuration and the aircraft assumes an angle of pitch P, the Doppler shifts measured for the two beams will be:

Forward beam $f_f = \dfrac{2F\,V_h\,\cos(\theta - P)}{c}$

Rearward beam $f_a = \dfrac{-2F\,V_h\cos(\theta + P)}{c}$

Therefore the difference is:

$$f_f - f_a = \frac{4F\,V_h\cos\theta\,\cos P}{c}$$

or when flying level:

$$f_f - f_a = \frac{4F\, V_h \cos\theta}{c}.$$

Port and starboard beams

When an aircraft is radiating two symmetrical forward beams, each subtending an angle A with the fore and aft axis of the aircraft, and the drift angle, or difference between the track and heading, is B then the Doppler shift associated with each of the forward beams will be:

Port beam $f_p = \dfrac{2\,F\,V\cos(A+B)}{c}$

Starboard beam $f_s = \dfrac{2\,F\,V\cos(A-B)}{c}$

where V is the true velocity along track.

The sum and difference between port and starboard Doppler shifts $(f_p \pm f_s)$ will provide the along axis velocity ($V \cos B$) and the across axis velocity ($V \sin B$) from which can be calculated V, the velocity along track (i.e. groundspeed), and B, the drift angle.

Type 80 Doppler navigation system

Typical of modern Doppler navigation equipment for smaller aircraft is the Type 80 navigation system produced by Racal Avionics Ltd. This comprises a Doppler radar equipment coupled to either a position Bearing and Distance Indicator (PBDI), automatic chart display, or directly into a navigational computer. For use with helicopters, it will also drive a special cross-pointer meter, known as a hovermeter, which will indicate low velocity along heading, lateral and vertical movement.

The Doppler velocity sensor

This consists of a transmitter, a receiver and an aerial. The transmitter module is a solid state continuous wave microwave source transmitting at a frequency of 13.325 GHz at a minimum power level of 30 mW.

The receiver uses a balanced mixer incorporating microstrip technology, an intermediate frequency amplifier using integrated circuits and suitable audio amplification. It is time-shared between the three receiving beams. Should the Doppler signal from any of the beams be lost, a suitable warning signal is sent to the computer/display. Built-in test equipment is incorporated, enabling the system to be checked at any time.

For transmission and reception, printed aerials are used which are mounted on the underside of the main transmitter-receiver assembly. Microwave switches in microstrip form produce a switched three-beam configuration. The depression angle of the beams is approximately 67° and the beamwidths are 5.5° in the depression plane and 11° in the broadside plane.

The Position Bearing and Distance Indicator (PBDI)

The PBDI accepts inputs from the Doppler sensor and a heading sensor such as a gyro compass. From this information it will derive and display the aircraft's position in latitude and longitude or grid coordinates. Alternatively, bearing and distance to a waypoint may be displayed. Up to ten such waypoints may be present, this information being retained even when the equipment is switched off. The display may also be switched to indicate groundspeed and drift angle.

The PBDI is now obsolescent although examples remain in service. It is now being replaced by navigation computers which also accept inputs from other sensors such as Omega/VLF, Loran C and GPS to provide full R-Nav capability.

Automatic Chart Display (ACD)

The ACD offers a simple means of indicating the aircraft's position on a standard chart. Position is indicated by the intersection of crossed wires positioned above the chart, the wires moving vertically and laterally in accordance with the computed position of the aircraft. If a wire reaches the edge of the chart, further movement of the display is inhibited but the navigational information is accumulated in the position stores until such time as either the aircraft returns within the area covered by the chart or the operator inserts a further chart.

Connections to the ACD are by flexible cable which enables it to be used either on the operator's knee or in any other convenient position.

The Type Doppler 80 equipment will supply heading velocity data to the computer in the range of 25 knots to 350 knots with an altitude range of 0 to 25 000 ft. The maximum heading error, assuming perfect heading input, is within 0.5% of distance flown along track and within 0.5° across track.

7.3 R-Nav

R-Nav, unlike the aids which we have previously discussed, is not a navigational aid with its own discrete sensors, but a computerised navigation management system accepting data from all available sources. Such sources may typically include: DME/DME, VOR/DME, INS/IRS, Omega/VLF, Loran-C, Doppler and Air Data from the aircraft instruments, with DME/DME normally being the preferred mode.

The data received from these sensors is then compared with a comprehensive database to give positional information in the form of latitude and longitude to the pilot. Comparison may also be made with a previously entered flight plan to indicate any course corrections necessary, or alternatively the equipment may be interfaced with the aircraft control systems to achieve this automatically. A comparison of positional information provided by the different sensors is also available to the pilot.

Plate 39 The controller for the Racal RNS 5000 R-Nav equipment. *(Photo: Racal)*

Distance to the next waypoint of the journey, together with the estimated time of arrival, may be displayed and, should it be desired to arrive by a given time, the R-Nav equipment will indicate the necessary Indicated Air Speed (IAS) or MACH to achieve this. If a hold is required at the waypoint, the automatic guidance will ensure that this is performed in accordance with published procedures. Typically up to 5,000 waypoints and 300 discrete routes may be stored within the navigation data bank of the system.

When operating in conjunction with ground based aids (such as DME/DME), the appropriate frequencies will be automatically selected and tuned. Should a required frequency be unavailable for any reason, the equipment will indicate this fact and make an alternative selection. In the event of a temporary total loss of data from the ground based aids, the equipment will maintain a Dead Reckoning (D/R) plot until data is again available.

In addition to navigation functions, the R-Nav may also provide assistance in fuel management by computing the present consumption and estimating fuel remaining at destination, maintain an engineering log by recording the nature and time of events, and assist in similar housekeeping duties.

Section 8

Space systems

8.1 Search and rescue satellites

By the mid-1970s, a number of countries had expressed interest in the concept of using satellite-aided SAR (Search And Rescue) systems and in 1975, Canadian scientists proved the feasibility of such a concept by means of a series of experiments using OSCAR (Orbiting Satellite Carrying Amateur Radio). This technique derived the position of the emergency beacon from the combination of an accurate knowledge of the orbit parameters and a measurement of the Doppler shift on the frequency of the relayed signal.

The following year, Canada, the United States and France began discussions on the possibility of an international programme and agreed three years later to the testing of an experimental SARSAT (Search And Rescue SATellite) system.

In 1980 the SARSAT partners joined with the Soviet Union for a joint demonstration project to be called COSPAS-SARSAT. This culminated in 1982 in the launch of the Soviet satellite COSPAS 1, the first to carry search and rescue equipment. With this, the North American ground stations also began operation. The following year COSPAS 2 reached orbit, shortly followed by SARSAT equipment carried on the United States weather satellite NOAA-E.

During this period, however, a number of other nations joined the project, including Norway, Sweden, United Kingdom, Finland and Bulgaria.

The viability of the system was proved only nine days after testing on COSPAS 1 began for, on 9 September 1982, the Ottawa ground station detected distress signals relayed by the satellite from an aircraft crash in northern British Columbia. The location co-ordinates provided by the system enabled searchers to find the aircraft in a mountain valley 90 km from its planned route. This aircraft was engaged in a search for the son of one of the occupants, whose aircraft was missing in the same region. Previously the Canadian SAR organisation had logged over 2,000 hours flight time in an unsuccessful search for the son, whilst the occupants of the COSPAS located aircraft were rescued within 24 hours.

Only a month later, on 10 October, the trimaran *Gonzo* encountered heavy seas and capsized 300 nautical miles south-east of Cape Cod. Fortunately an EPIRB was carried and although the distress signals were heard by transatlantic aircraft, it was not until COSPAS-SARSAT data was available that the US Coastguard was able to locate and rescue the three survivors.

COSPAS-SARSAT helped to save 28 lives during its first six months in operation.

Operation

In simplest terms, the principle of the COSPAS-SARSAT system is that a series of satellites in low polar orbit listen for distress signals. When received, these are relayed to a Local User Terminal (LUT) where, using a measurement of Doppler shift on the frequency of transmission and a precise knowledge of the satellite orbit, the origin of the distress signal can be derived. This information is passed to the Mission Co-ordination Centre (MCC), which in turn alerts the Rescue Co-ordination Centre (RCC), which instigates the search and rescue operation in accordance with normal practice.

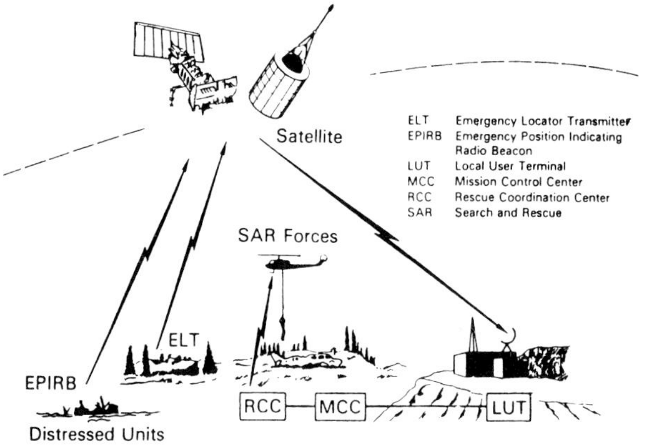

Fig. 58 The COSPAS-SARSAT basic concept

Limitations

Currently the satellites are operating in conjunction with the present generation of ELTs (Emergency Locator Transmitters) and EPIRBs (Emergency Portable Indicating Radio Beacons) with COSPAS listening on 121.5 MHz and SARSAT on 121.5 and 243 MHz. Equipment operating on these frequencies, however, was not designed for detection by satellites, and this causes severe problems, not the least of which is the low transmission power – often in the region of 50–100 milliwatts.

To broaden the capabilities of the COSPAS-SARSAT system, a new generation of ELTs and EPIRBs operating on 406 MHz (a frequency for which the current satellites are also equipped) is being developed.

This new equipment will give considerable advantages, the first of which is global coverage. The 121.5 MHz system operates in real time, requiring the emergency beacon and the LUT to be simultaneously visible from the satellite. This restriction will not apply on the 406 MHz channel.

As a consequence of the higher frequency, higher positional accuracy will be possible. Furthermore, each beacon will carry unique identification and the capability of transmitting a digitally coded message such as details of the distress situation and (if known) latitude and longitude.

Results

The results from the project so far have been a major success. Despite an extremely high false alarm rate (as much as 98% in North America), between September 1982 and May 1984, the system provided alert and location data for 85 distress incidents world-wide involving 237 persons, from which 214 survivors were rescued.

The accuracy, too, has been remarkable. Working on a database of nearly two hundred cases in which the ELT was located, the mean radial error was less than nine miles and over 60% of detections were within twelve miles.

In view of the demonstrated success of the system, the COSPAS-SARSAT nations have agreed to provide search and rescue satellite services at least until 1990. The system is to comprise four satellites, two being provided by the Soviet Union and two by the SARSAT countries.

The combined COSPAS-SARSAT system has now become such an important tool in search and rescue that the associated nations have further decided to develop an institutional basis for the organisation, management, funding, and future development of the system.

8.2 Geostationary meteorological satellites

Meteorological satellites fall into two main categories: low-level orbiting (such as the NOAA series), and geostationary, whose orbit is above the Equator and whose period is exactly the same as the rotation of the earth and thus always remains at exactly the same point above the earth.

A series of five equi-spaced geostationary satellites are in orbit, all operating in a similar manner. The satellite positioned over South America and another over the Pacific Ocean were provided by the USA, the one over the Indian Ocean by he USSR, the one over New Guinea by Japan, and last, Meteosat, just south of Ghana, by Europe.

Meteosat has been designed to carry out three main missions: Image Production, Data Collection and Information Dissemination.

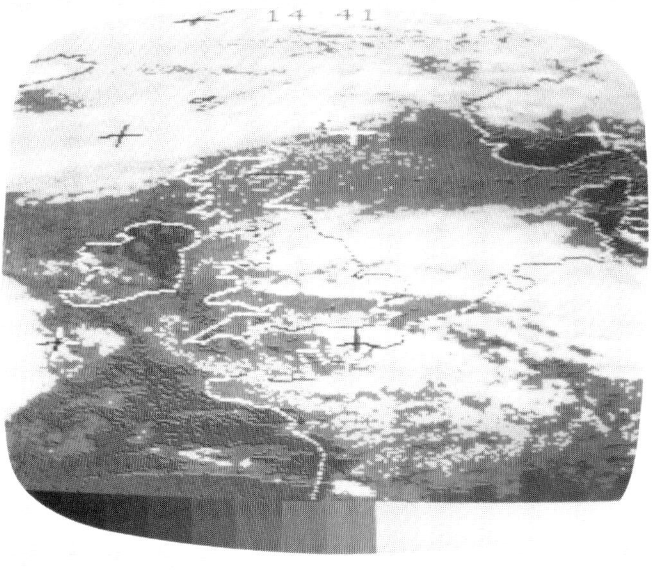

Plate 40 A typical Meteostat processed image, showing the UK and parts of Europe. *(Photo: Feedback Instruments Ltd)*

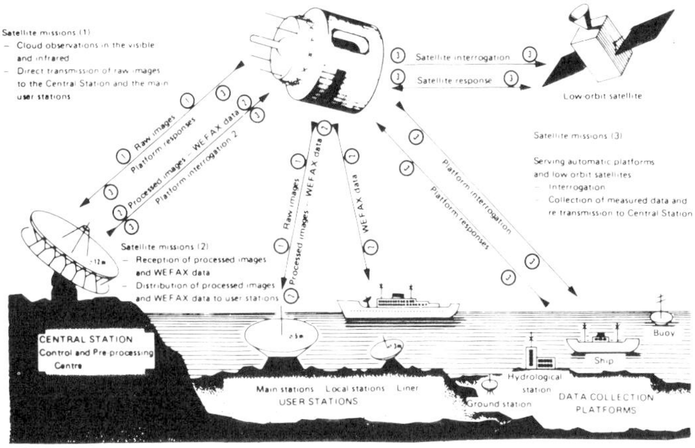

Fig. 59 The overall Meteosat mission, illustrating the types of data transmitted to and from the satellite.

The first of these involves the scanning, at half-hourly intervals, of the earth's surface and cloud masses within the coverage area, in three spectral bands. Secondly, Meteosat collects information gathered by automatic or semi-automatic stations (called Data Collection Platforms), or by other satellites in low polar orbit, the purpose being to gather information obtained locally to supplement that obtained as part of the main mission. The third function is to disseminate the cloud cover images and meteorological data derived from these images, in order to ensure that the largest possible number of users have access to the data produced by Meteosat and the Data Collection Platforms.

The spacecraft

Meteosat is spin stabilised at 100 rpm on an axis perpendicular to the orbital plane. It is of cylindrical shape, 3.2 metres high and 2.1 metres in diameter. At launch, the mass was nearly 700 kg, including the apogee motor and the securing devices.

The design is relatively simple, involving the use of a double structure. The primary structure bears, in addition to the mechanical loads, a main platform carrying the support equipment. The upper platform carries the aerials and most of the communications equipment. The secondary structure supports the six solar panels and the heat shields.

In order to achieve the design task, the payload consists of a high resolution radiometer and a data transmission system. The radiometer is an electro-optical instrument whose main element is a 40 cm aperture Ritchey-Chrétien telescope. This can, over a period of 25 minutes, produce two simultaneous images of the earth's surface, one in the visible and the other in the thermal infra-red regions of the spectrum. An infra-red water vapour channel is also available.

Tracking and processing

The Data Acquisition, Telecommand and Tracking Station (DATTS) is situated at Michelstadt near to the European Space Operations Centre (ESOC) at Darmstadt in the Federal Republic of Germany. This facility is responsible for the reception of the radiometric, attitude and housekeeping data from the satellite. In addition it transmits the meteorological data or images and the telecommands and finally, in association with a land-based transponder, determines the position of the satellite by use of ranging techniques.

The operational management of the system is the responsibility of the Meteosat Operations Control Centre (MOCC). This involves monitoring the operation and performance of all elements within the system, including performing the orbit and attitude restitution calculations.

Signal processing and the production of the images is the task of the Data Referencing and Conditioning Centre (DRCC) and from the processed data, the Meteorological Information Extraction Centre (MIEC) extracts the specifically meteorological data such as wind fields, sea temperature charts and cloud system analyses.

These last three centres, MOCC, DRCC and MIEC, are actually utilisation consoles connected with the Meteosat Ground Computer System (MGCS), a computer unit located at ESOC and linked with DATTS by a high-speed terrestrial circuit.

From its geostationary position in orbit the satellite observes the cloud cover of the earth, producing two images in 25 minutes. The first of these is of visible radiation and is composed of 5,000 lines each of 5,000 image points and achieving a resolution of 2.5 km at the earth's surface. The second, in the infra-red spectrum, comprises 2,500 lines, each with 2,500 image points giving 5 km resolution. When received on the ground, these images are of sufficiently high resolution for immediate use, but for more precise definition and for the data to be compatible with international APT (Automatic Picture Transmission) or WEFAX (Weather Facsimile) standards, the raw image must be processed on the ground and then re-transmitted to the satellite for relay to the weather stations.

The radiometer

The radiometer consists of a main optical unit and some additional electronic packages mounted on the satellite equipment platform. The main outward feature distinguishing the European radiometer from similar equipment developed in the United States is the absence of a scanning mirror as the first element in the optical chain. To avoid the need for this, the Meteosat design scans the primary telescope. This consists of a Ritchey-Chrétien primary and secondary mirror mounted with a small 45° on-axis mirror in a Condé-Cassegrain arrangement. By use of this system the telescope may be scanned about a plane which nominally lies in the satellite's, and hence the earth's, equatorial plane whilst still maintaining the subsequent relay optics, detectors, etc. in a fixed position with respect to the satellite.

On exiting the telescope, the optical axis is folded by a series of small flat mirrors to arrive either at the optical detectors or via a lens system to the infra-red detectors.

To maintain the necessary optical quality over a wide range of temperatures and in the presence of significant thermal gradiants, the mirrors are fabricated from a low expansion material from which material in excess of structural requirements has been removed by ultrasonic drilling and grinding processes.

In normal use the frame (vertical) scan of the image is generated by rotating the telescope through ±9° in 2,500 steps, once per spacecraft revolution. As the telescope is rigidly fixed in the other plane, the linescan is developed by the rotation of the satellite.

The optically collected visible and infra-red earth signals are converted into analogue electrical signals by five detectors, two visible and three infra-red. The two visible detectors are located in the focal plane of the primary telescope, both being fabricated on a single silicon chip. This ensures the homogeneity of performance required for good image quality as each detector generates alternate lines of the high resolution image.

Whilst the visible detectors operate at ambient temperatures, the three infra-red detectors, i.e. a redundant pair for thermal infra-red and a single element for atmospheric water vapour images, must be cooled to less than 95°K. These are therefore insulated in a second stage, or cold patch, of a passive cooling system which uses what is effectively a black body radiator mounted at one end of, and thermally insulated from, the spacecraft. Even this is insufficient to completely remove the sun's influence on detector temperature the whole year round. Active heating is therefore employed to keep the detector operating temperature constant and stabilise the otherwise temperature-variable spectral-sensitivity characteristics.

A further facility of the main unit is that for in-flight infra-red calibration with an associated black body reference source. This mechanism permits three calibration conditions: detector self viewing, detector sun viewing, and detector black body viewing. Viewing operations, the moon, space background and selected earth regions, can also be used for referencing but this is due to the inherent flexibility of the equipment rather than any special on-board equipment.

The synchronisation sub-system of the imaging system generates all the dating and control signals. This system is controlled by a crystal oscillator operating on 5.3 MHz which is reset by a sun (or earth) sensor for each orbital revolution of the satellite. From this is synchronised the radiometer scanning, the despun aerial and the transmission of data to earth.

The analogue data from the radiometer is converted into digital signals at high velocity by image processing units and a memory circuit provides temporary storage in order that the data can be transmitted at a slower, continuous rate.

Telecommunications

The main telecommunications sub-system is responsible for the transmission to earth of the data produced by the imaging system and the re-transmission of the processed images to the various user stations, either in digitial form for higher resolution images or in analogue form compatible with APT or WEFAX formats.

Links between the satellite and ground stations operate on S-Band, except those between the spacecraft and Data Collection Platforms which operate on UHF, in which case the satellite is operating as a UHF/S-Band repeater.

Separate aerials are used for S-Band transmission and reception. The signals are radiated from a directional electronically despun aerial which has a gain of approximately 13 dB. This comprises an array of 128 dipoles arranged in 32 rows of 4 on a cylindrical structure which is fed from a switching matrix. This energises the 5 appropriate rows of dipoles at any time in synchronism with the satellite rotation. For reception a 2.5 dB gain toroidal pattern aerial is used with a further aerial of the same type provided as a back-up to the transmission arrays. Each of these consists of a slotted waveguide array with linear polarisation. They are placed side-by-side within a cylindrical honeycomb radome. UHF links are achieved by a further toroidal array of unity gain consisting of four half-wave dipoles etched on a microwave plastic cylinder.

During the stationing phase, VHF is used for telemetry and command. The aerial for this comprises four monopoles fed by signals of equal amplitude in phase quadrature.

Receiving the images

Although, due to the distance from earth and the scanning method used, the whole earth is continuously viewed by the satellite, the images received by the user are vastly different, for on normal reception channels the satellite is operating as a repeater, relaying only sections of the whole earth picture. These have been processed at the ground facility, have coastlines and reference markers added and are radiated to a published schedule. By radiating only a segment of the earth at any one time, greater detail is visible on the monitor and some equipment has the facility for a further magnification.

Receiving other satellites

The five geostationary meteorological satellites are augmented by a number of others in low-level polar orbit such as the NOAA series. These use the same signal format as Meteosat and consequently, if a suitable aerial is connected directly to a VHF receiver, may be received when above the horizon.

8.3 Global Positioning Systems

There has always been a requirement for a world-wide navigational aid with an accuracy of a few metres. With the introduction of the Navstar Global Positioning System (GPS) and Glonass, it would appear that this ideal has been very closely approached.

The basic principle of both systems is similar: namely that a number of satellites in orbit each radiate a series of precisely timed radio signals. The user notes the time at which the signals are received and from the delay of each due to transit time and knowledge of the position of the satellite at the moment of transmission, calculates his distance from each and thus his position.

Simple though this basic concept may be, in practice difficulties abound and in their solution, two navigational aids have been developed which are capable of resolving the user's position on earth to nearer than 100 metres for civilian equipment and approaching an order better for military-survey purposes. It is expected that they will remain in service well into the 21st Century and development during that period will result in even higher accuracy.

Navstar

The development of the Navstar GPS was funded by the United States Dept of Defense and began in 1973 when the United States Air Force, Army, Marine Corps, Navy and Defense Mapping Agency combined their technical resources to develop a highly accurate space-based navigational system.

The space segment

The space segment will ultimately comprise eighteen satellites in six separate orbital planes plus three satellites for operational back-up.

For navigational purposes, each satellite radiates on two frequencies in L-Band; 1575.42 and 1227.6 MHz, these being known as the L1 and L2 signals respectively. Two modulations are used: Precise (P), intended for military-survey applications, and Coarse-Acquisition (C/A) for general use. The L1 signal will carry both modulations but L2 will carry either P or C/A but not both.

Four further frequencies are carried on each satellite: two S-Band channels for refreshing the pre-calculated spacecraft position memories and correcting the satellite clock from ground stations, and a further L-Band and a UHF channel for the spacecraft's second payload. The signals are radiated from a shaped beam aerial which provides a power level of at least −166 dBW to users.

Navigational information is simultaneously provided on two frequencies in order that inaccuracies due to refraction effects in the ionosphere may be eliminated, these being approximately inversely proportional to the square of the frequency.

Each Navstar satellite has a weight of 850 kg (1,862 lb). It orbits the earth at a height of 10,898 nautical miles with a period of 12 hours in one of six orbital planes each inclined at 55° to the Equator. It is expected that each satellite will have an operational life of about seven and a half years.

The information channels

The C/A signal is a pseudo-random but predetermined digital signal, unique to each satellite, of 1,023 bits clocked at a rate of 1.023 Mbps, which repeats continuously. Each sequence therefore occupies one millisecond.

In addition to the pseudo-random code, the satellite also radiates a data bit stream at 50 bps. This provides information on the status of the space vehicle, the time synchronisation information for transfer from the C/A to the P code, the parameters for computing the clock correction, the ephemeris of the space vehicle, and the corrections for delays in the propagation of the signal through the atmosphere. In addition, it contains almanac information which defines the ephemerides and the status of all the other space vehicles, this being required for use in signal acquisitions. The data format also includes provisions for special messages.

The navigation message is formatted in five subframes each six seconds in length, which combine to make a complete data frame of 30 seconds, 1,500 bits long. The precision (or P) code is also pseudo-random, but the data rate is ten times higher, at 10.23 Mbps. This is intended for military and survey purposes, will not be used in normal civilian applications, and in consequence will not be discussed further here.

The precision timing of the transmissions of each satellite is controlled by an on-board atomic clock of 10.23 MHz nominal frequency, whose accuracy is such that it would gain or lose only one second every 36,000 years. Even this, however, is insufficient, so the space vehicle timing is compared with caesium clocks maintained by the Master Control Station (MCS). Corrections are transmitted to the satellite where they are relayed to users via the 50 bps data stream.

The pseudo-random codes, which are synchronised to the space vehicle time, are maintained within 976 microseconds of the GPS system time in order to preclude secondary control problems, such as almanac word length limitation, which would otherwise arise.

All frequencies in the satellite are derived from and synchronised to integrals of the 10.23 MHz frequency standard. These include:

	Repeat interval or frequency
P-code	
reset	7 days
frequency	10.23 MHz
C/A code	
Epoch	1 millisecond
frequency	1.023 MHz
L1 RF frequency	154 × 10.23 = 1575.42 MHz
L2 RF frequency	120 × 10.23 = 1227.6 MHz

Resolving the signal

In a perfect world, the determination of the distance of the spacecraft would be simple – merely compare the time of arrival of the signal with the user equipment internal clock and, knowing the speed of transmission of electromagnetic radiation, derive the distance between user and spacecraft. Repeat the process with two or three other satellites and from these results derive the user's position.

Unfortunately, life is not quite so simple, so techniques have to be developed to allow for the real-world situation. The first task of the user's receiver is to synchronise with the received satellite transmission. This is achieved by generating an internal pseudo-random code, identical to that radiated by the satellite in use. This is then compared with the received signal and when correlation is obtained, the time of receipt of the signal can be determined.

This, however, could be subject to considerable ambiguity, for the C/A code epoch is just one millisecond. As the speed of electromagnetic radiation is 186,000 miles per second, this corresponds to a distance of 186 miles per millisecond. However, the satellite is positioned in a high orbit and in general the spacecraft-to-user distance will be in excess of 11,000 miles – a transit time in the order of 60 milliseconds. This ambiguity is resolved by inverting the phase of the stream at 50 Hz, i.e. every 20 milliseconds. Some ambiguity will still exist, but it would be hoped that the user has some idea of his position – certainly within 3,000 miles!

With the C/A signal synchronised within the receiver and the ambiguity resolved, the 1.023 MHz pseudo-random signal may be used as markers to interpolate between the 1 millisecond (186 mile) repetitions, i.e. to approximately 960 ft. Further interpolation within the set will provide even greater accuracy. If the distance to two other satellites is then determined, the user's position may then be derived.

Such accuracy of distance determination is, however, dependent upon very accurate synchronisation between the spacecraft and user equipment clocks, for every nanosecond discrepancy will result in a 1 ft error.

Unfortunately, the cost of inclusion of a clock of such accuracy within the user equipment would make the installation of GPS equipment totally uneconomic. Another means of deriving the GPS time had therefore to be sought and this was achieved by the use of a fourth satellite for obtaining a positional fix.

In understanding this technique, let us first consider that the user is at a position described by the co-ordinates Ux, Uy, Uz, and the satellite is at range R1 at position co-ordinates X1, Y1, Z1. The clock bias (i.e. the discrepancy between user clock and GPS time) = CB.

Then:

$$(X1-Ux)^2 + (Y1-Uy)^2 + (Z1-Uz)^2 = (R1-CB)^2 \quad (1)$$

and for the second, third and fourth satellites:

$$(X2-Ux)^2 + (Y2-Uy)^2 + (Z2-Uz)^2 = (R2-CB)^2 \quad (2)$$
$$(X3-Ux)^2 + (Y3-Uy)^2 + (Z3-Uz)^2 = (R3-CB)^2 \quad (3)$$
$$(X4-Ux)^2 + (Y4-Uy)^2 + (Z4-Uz)^2 = (R4-CB)^2 \quad (4)$$

In these equations, the values for X1, Y1, Z1, X2, Y2, Z2, X3, Y3, Z3 and X4, Y4, Z4 are provided by the satellites and R1, R2, R3, R4 are measured times. This leaves only Ux, Uy, Uz (the user's position) and CB (the clock bias) unknown. With four equations, solutions for all unknowns are possible, this being achieved within the receiver using iterative techniques.

In order to maintain the full possible accuracy of the Navstar GPS, the satellites are monitored and updated at least once per day by the Master Control Station. These updates include clock correction factors, satellite ephemeris constants (orbital elements) and information on the current status of the earth's ionosphere. Corrections are also made for relativity effects in accordance with Einstein's special and general theories of relativity. Due to this, for example, allowance is made for the fact that the user equipment is affected by stronger gravity than is experienced by those on the satellite. For this, the satellite clock frequency, nominally 10.23 MHz, is offset to 10.2299999945 MHz.

Present experience with the Navstar GPS indicates that, despite the constellation of satellites being incomplete, an accuracy of about 15 metres is possible using the C/A code, and about 6 metres with the P code. However, as Navstar is essentially a military system, albeit readily available for civilian use, the capability of considerably reducing the system accuracy for other than authorised users in times of national emergency has been incorporated into the design.

Glonass

While the United States was developing the Navstar GPS system, the Soviet Union was developing an equivalent system for similar purposes. It is not surprising, therefore, that there are remarkable similarities between the two systems.

The complete constellation of space vehicles will comprise 24 satellites in

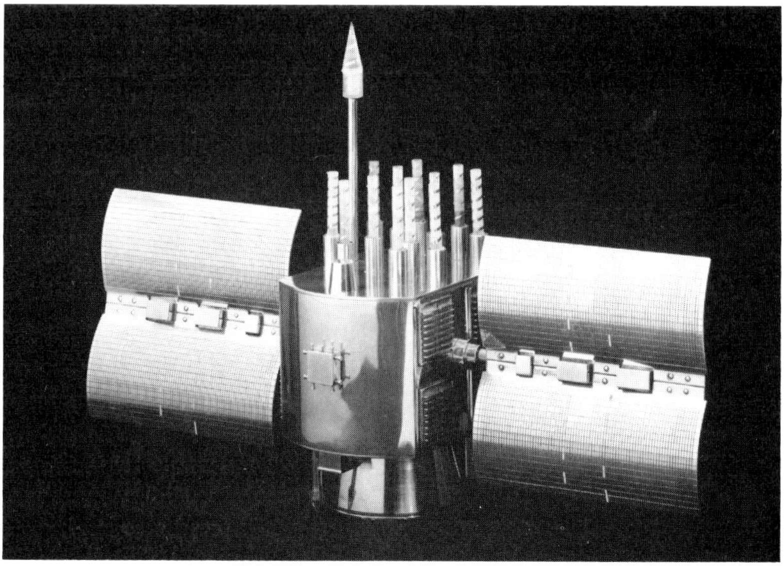

Plate 41 The Navstar satellite. *(Photo: Plessey)*

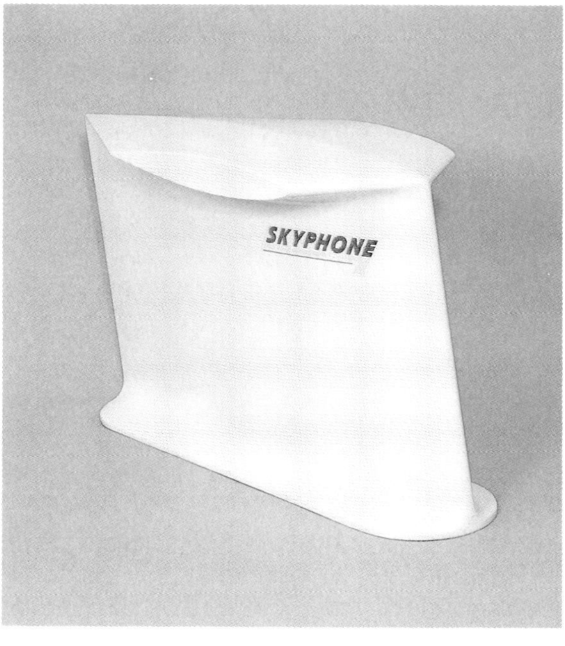

Plate 42 Typical high gain aircraft antenna for satellite communication. *(Photo: Racal Antennas Ltd)*

three orbital planes at an altitude between 18840 and 19940 km with an orbital period of 11 hours, 16 minutes. The satellites each radiate L1 signals at approximately 1600 MHz and L2 signals at approximately 1250 MHz. However, the method of identification is different, for whilst Navstar satellites all transmit on the same frequency and identify by means of their pseudo-random signal, the Glonass satellites each radiate on a discrete frequency, the L1 frequency being 9/7 times that of the L2. The channel spacing is 0.5625 MHz on L1 and 0.4375 MHz on L2 frequencies.

Glonass satellites radiate a wideband P code on both L1 and L2 frequencies with the C/A code on L1; however, the bandwidth used is almost exactly a half of that radiated by Navstar, being ± 5.11 MHz for P code and ± 0.511 MHz for C/A.

The data frame comprises 7500 bits transmitted over a period of 150 seconds. This is radiated as five subframes, each of 1500 bits divided into 15 words.

The navigation message transmitted from each satellite contains both ephemeris data and almanac. The former relates only to the transmitting satellite and includes: three coordinates; three components of velocity; and three components of acceleration caused by the earth and moon gravity at the defined time points; satellite time marks; clock bias relative to system clock; and clock drift relative to the system reference frequency.

The system almanac provides operational status data of the satellites of the system, coarse values of each satellite clock bias relative to the system clock, orbital parameters of all satellites and system clock bias relative to UTC.

Although both Glonass and Navstar were nominally referenced to UTC, there was originally a variation between the two of some 27 microseconds. However, in recent years this difference has been reduced and at the time of writing (early 1992) is virtually zero.

A final major difference between Navstar and Glonass is that whilst the former system has the capability of degrading accuracy for non-authorised users, this has not been provided in the Glonass system. It is possible, therefore, that in future the latter system may be preferred by civilian users on the grounds that the accuracy of the system may be relied upon despite any change in international relations.

The future of satellite navigation systems

It would appear inevitable that, given the accuracy and universal capability of satellite navigation systems, except where extreme short range accuracy is required, these will eventually supersede the plethora of ground-based aids currently in use. How soon this will occur will depend largely on their availability and cost.

In its simplest form the information display from a Sat-Nav receiver will give the latitude and longitude of the user, but the combination of this with various databases is already available.

In the maritime environment, positional displays are available which present a map of a given area with the user's position indicated by a bright dot. The map can indicate a relatively large area, such as the North Sea, yet the area viewed can be 'zoomed in' to give fine detail such as individual jetties at ports and harbours. Within a short timescale it is expected that sufficient memory will be available to present world-wide maps without recourse to external memory cartridges, etc.

Additionally within this system, numbers of waypoints can be programmed into the equipment with the capability of providing the necessary navigational instructions (course, speed, etc.) to complete a given journey in the most economical manner.

These facilities are available at a price which makes them available to relatively unsophisticated privately owned boats.

Such facilities are currently available to aircraft using R-Nav equipment in conjunction with ground-based aids and it is inconceivable that this will not be extended within the very near future to accept GPS input with the addition of suitable map displays.

At this point it is within the bounds of possibility that GPS systems may well render many of the en-route navigational aids, and even the CAT. 1 instrument landing systems common in many parts of the world, unnecessary.

8.4 Satellite Communication Systems

One of the major problems of long haul aircraft operations has been that on trans-ocean routes, the only communications available are via HF SSB radio links which are less reliable and frequently of lower quality than may be desired.

However, with the development of satellite communications and in particular the INMARSAT system of geostationary communications satellites, reliable links are at last possible.

As at early 1992 these were still in their infancy, and the initial installations have provided only for telephonic communication for passengers. Future developments are expected to provide air traffic control communications and proposals exist for the aircraft position, as provided by the on-board navigation system, to be transmitted to the appropriate ATCC. This could then be indicated on a 'pseudo-radar' type display for air traffic control officers. The effect of this could lead to reduced aircraft separation standards with consequent higher traffic density capability.

The aircraft system

The user interface in the aircraft, which could comprise the flight crew headset, the navigational computer or a passenger handset, is coupled to a satellite data unit. This unit digitises the speech waveform at 9600 bits/second, adds an aircraft technical address for identification purposes and controls voice and data coding, system protocol and timing functions. The signal is then coupled to a radio frequency unit which converts it to L-band and makes any necessary Doppler corrections. The signal then passes to the high power amplifier and then, via a beam steering unit, to the antenna.

The antenna may typically comprise three electronically steerable phased arrays which cover the port, starboard and zenith regions. These are contained in a single blade type radome designed for top mounting on the aircraft fuselage. The radome contains all phase shifters and power splitters necessary for beam steering. The gain would be in the order of 12 dBic over a coverage zone of 360° in azimuth and from +5° to +90° in elevation.

The antenna would also provide a discrimination of 13 dB against the signal from a satellite spaced 45° or more from the wanted satellite.

The space segment

On reaching the satellite, the frequency is again converted, this time to the

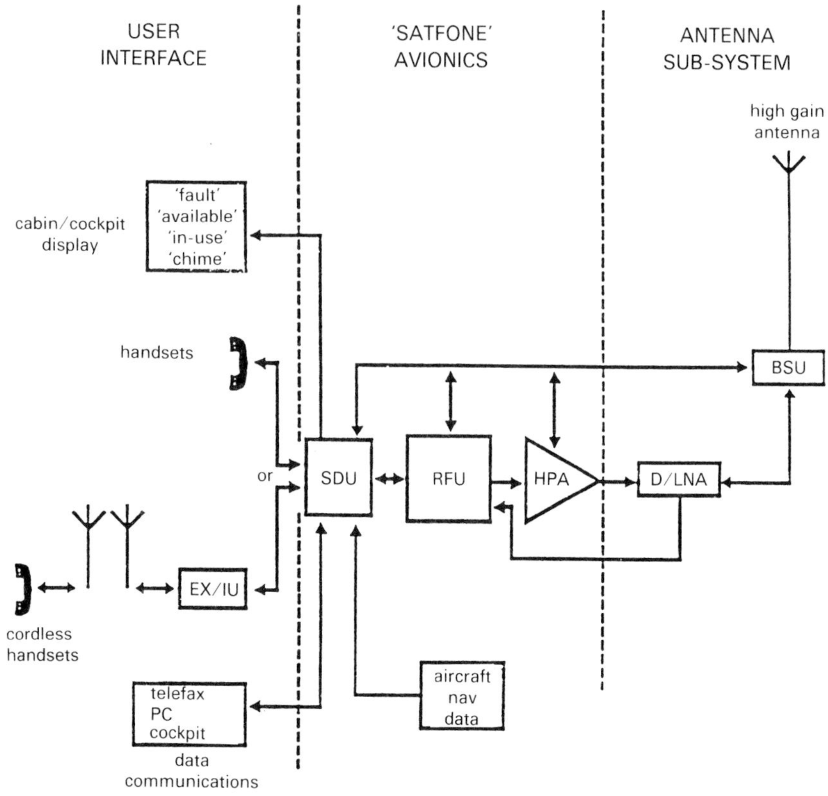

Fig. 60 Schematic of the aircraft satellite communications installation. SDU, Satellite Data Unit; RFU, Radio Frequency Unit; HPA, High Power Amplifier; D/LNA, Diplexer/Low Noise Amplifier; BSU, Beam Steering Unit; EX/IU, Exchange/Interface Unit.

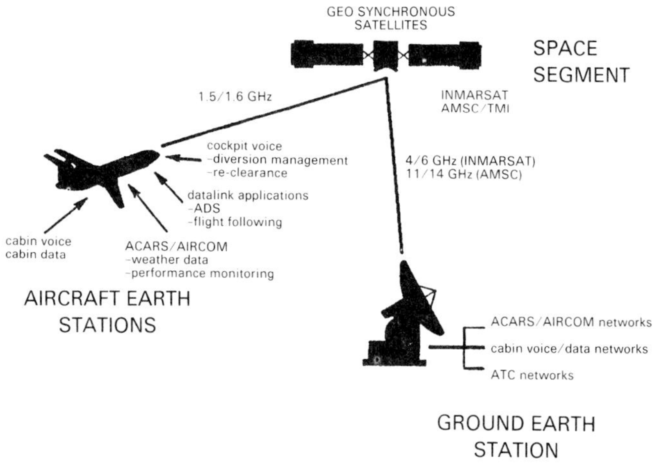

Fig. 61 Overall schematic of satellite communication. (Diagram: Racal Avionics)

4/6GHz waveband for transmission to the appropriate ground earth station from where it is connected to existing public and private telephone networks.

Inmarsat operate a constellation of geostationary satellites which will provide world-wide coverage below a latitude of 80° North and South. Each of the first generation satellites (Inmarsat-1) provides 40 channels to aero-satcom users; however, the Inmarsat-2 satellites currently being deployed provide 200 channels for this function. Although 200 channels per satellite may seem small, analysis indicates that the possibilities of all channels being used simultaneously are small. Nevertheless, the third-generation Inmarsat constellation of four satellites is expected to be launched in the mid-1990s, each of which will provide in the order of 2000 communications channels. These satellites will also offer 'spot beam' capability which will enable the re-use of the same frequencies in two separate areas at the same time.

Section 9

Miscellaneous systems

9.1 Runway visual range assessment

Before discussing the automatic assessment of Runway Visual Range (RVR) it is essential that the basic problems concerned with RVR assessment in general be thoroughly explored.

The need for accurate RVR assessment

There are two reasons why an assessment of RVR is required, one operational and the other regulatory. In the first instance, before attempting an approach to a runway for landing, the pilot of an aircraft needs to know whether the visibility on the final stages of that approach and along the runway is adequate to permit a safe landing to be achieved.

Complementary to the pilot's requirement, aircraft and runways are categorised according to the approach facilities with which they are equipped. The worst conditions under which an approach may be attempted is defined in terms of cloud base and RVR. To enable such regulations to be applied, both these factors must be measured to some considerable degree of accuracy, particularly at the times when weather criteria are changing from one category to another.

When such transition points are reached in a deteriorating meteorological situation, whole groups of aircraft are excluded from even attempting a landing.

From the foregoing it is evident that a perfect RVR measurement would meet both pilots' and regulatory criteria. RVR measurement is, however, of such complexity that it can never be considered other than an assessment.

As the overriding requirement of the regulatory function is safety, it is only necessary for this purpose to apply a sufficiently large factor to allow for error and any method of assessment may be used provided that the magnitude and probability of error is known. Although such means will not provide a correct assessment of RVR, nevertheless, any errors will not mislead an approaching pilot into a hazardous situation. The very nature of such assessments can mean that if significant errors exist, they will be on the 'safe' side and consequently it is possible that an approaching pilot may be refused the opportunity to land when it is perfectly safe for him to do so. These factors have led to considerable effort being expended on the development and provision of automatic systems for accurate assessment of RVR to enable the requirements of safety to be met whilst encouraging the expeditious flow of air traffic.

The assessment of RVR

The traditional method of assessing RVR was by positioning a caravan or hut adjacent to the runway from which an observer could gauge the visibility by observing marker boards adjacent to the runway in daytime and runway lights at night. In earlier times this could give a highly representive assessment for the eye of the observer approximated very closely to that of the pilot on the runway.

Modern conditions have reduced the efficiency of such methods, for apart from the increased height of modern aircraft, present-day restrictions on obstacles adjacent to the runway necessitate such observations being made from a point at such a distance from the runway that the assessment is not necessarily representative of runway conditions.

Other factors also reduce the reliability of the human observer. When visibility is low the identification of particular lights becomes more difficult and if a light is not seen it may be due to either reduced visibility or lamp failure. The adaptation of the eye at night presents further problems. After leaving a well-lit environment, adaptation may take up to half an hour, yet, even under optimum conditions, a glance at the brighter lights is sufficient to impair the eye for several minutes and more distant lights may remain invisible until adaptation is again complete.

In evening conditions, if the light is fading faster than the eye adapts, erroneous assessment can occur. When the observer is positioned in a hut and observing through a window, dirt, condensation and misting can reduce the apparent visibility. The lights themselves also suffer these factors and, in addition, the intensity of their output reduces as light bulbs age.

Despite all these factors, however, an observer assessment of RVR is quite valid provided that due allowance is made for possible errors.

Although in many countries RVR assessment is still made by human observer, the expansion of the observer system is not practicable because of the rapid, simultaneous reporting from a number of observation posts required by ICAO recommendations.

There are three of these. Firstly, that RVR should be reported at a rate of between once per second and once per minute. Secondly, that independent reports should come from different parts of the runway. Thirdly, that series of fixed values of RVR should be used for reporting, the size of step from one value to the next being:

(a) Below 200 m in 25 m increments
(b) Between 200 m and 800 m in 50 m increments
(c) Above 800 m in 100 m increments

In order to meet these requirements a move towards automatic assessment had to be made. In 1969 the Telecommunications Directorate of the National Air Traffic Services issued a specification for a fully automatic RVR system to meet the ICAO requirement. Within the UK, an equipment was developed to meet this specification which was thoroughly evaluated in the winters of 1970–71 and 1971–72 before being introduced into full operational service. Instrumented Runway Visual Range (IRVR) systems have now been developed in many countries, but a description of the British system will provide an excellent example of the principles involved.

In the automatic assessment of RVR, the first major problem is automating the function of the observer. Notwithstanding the problems of using manual techniques to assess RVR, the observer method does at least use the same sensor as the pilot–the human eye. Thus, if the quite reasonable assumption is made that all healthy eyes are much alike, the special characteristics of the eye can be ignored. In replacing the human eye with an alternative form of optical sensor it is essential in using that sensor that the essential characteristics of the human eye are retained. Certain aspects of human vision must therefore be quantified before visibility can be assessed by machine or in other words a standard eye must be derived before it is possible to define what that eye can see.

In the definition of a standard eye, consideration must first be given to the limitations of human vision which can be measured in an individual and from which a mean can be obtained for population sample.

There are three such limitations: the threshold of contrast, the threshold of illuminance and the threshold of dazzle. It may well be argued that all are manifestations of the same effect but for practical purposes visual range assessments are gauged in terms of visual threshold and contrast threshold.

Contrast and illuminance thresholds

If an observer, with eyes fully adapted to the dark, looks into absolute darkness, nothing is seen except occasional small scintillations of light which are products of the retinal nervous system. If the absolute darkness is then relieved by a uniform diffuse light of gradually increasing intensity, a point will be reached at which the observer first notes that it is not quite dark. This level is the threshold of perception of light, or illuminance threshold.

If the light level is increased still further, nothing more is noted except the increased light intensity. However, if the light level in just one part of the observed area is increased independently, then at some level of illumination that area becomes visible by contrast. When this occurs the contrast may be defined as the brightness contrast threshold of the observer's eye.

In bright light, objects are seen by contrast either of colour or brightness. In a well-lit scene all objects emit, or reflect, light at a level above the visual threshold of the observer's eye. In these circumstances, to make an object less visible it is only necessary to make it the same colour and brightness as its background. Conversely, to make the object more evident, its brightness level must be enhanced compared with the background and/or its colour must be changed.

If the illumination of the scene is progressively reduced, at some instant most of the scene is reflecting light at a level which is below the observer's visual threshold. The remaining objects are then visible by stark contrast against the black background and they will remain visible until their illumination level drops below the illuminance threshold. This happens at night when distant lights are seen by stark contrast until their level falls so low as to produce no detectable energy in the observer's eye. The contrast and illuminance thresholds of a typical eye have been established and agreed internationally for the purpose of assessing visibility.

Distant vision

The foregoing paragraphs have assumed that the scene viewed is not either partially or wholly obscured by anything between it and the observer. This is a valid assumption in clear conditions over normal distances but the scene which is in sharp contrast when viewed from close range becomes indistinct as the observer retreats. There are four main reasons for this effect:

(a) As objects retreat they subtend a smaller angle at the eye until they reduce below the definition of the eye.

(b) The light emitted or reflected from the objects reduces with distance in accordance with the inverse square law, thus reducing contrast.

(c) Light is lost by absorption, or scattering, in the atmosphere along the path.

(d) Incident light being scattered back into the observer's eye reduces contrast.

The effects of the atmosphere i.e. scattering and absorption, have been investigated and as early as 1876, Allard formulated a relationship between the limit of visual detection of lights and the illuminance threshold of the eye, whilst in 1924 Koschmieder derived a law relating contrast threshold to visual range.

Koshmieder's law

In viewing unlit objects in daytime the eye appreciates visibility as a problem of recognition of contrast between the object and its background. Unfavourable atmospheric conditions reduce this contrast; in fact, losses in transmission through the atmosphere reduce the available contrast by the amount of the loss.

In flying operations, as in meteorology, the minimum detectable contrast is taken as 5% to allow for poor initial contrast such as painted markings on wet concrete.

The law derived by Koshmieder was:

$$E_c = T^{R/Z} .$$

where
E_c = contact threshold,
T = atmospheric transmission over a baseline Z,
R = visual range.

Allard's law

At night, and in the majority of daytime fogs, runway lights can be seen by the pilot before the runway or its markings. The limit of visual detection of lights is set by the illuminance threshold of the eye according to the relationship which was formulated by Allard in 1876:

$$E_t = \frac{I}{R^2} T^{R/Z}$$

where
E_t = illuminance threshold,
I = intensity of lights.
Whereas the contrast threshold, E_c in Koshmieder's law, is independent of the absolute level of background brightness, the illuminance threshold E_t is directly related to it.

The relationship has been determined empirically a number of times, in varying circumstances; the one generally adopted by international bodies is derived from the work of Blackwell as published by the *Journal of Optical Societies (America)* in 1946 and is:

$$\log E_t = 6.95 + 0.887 \log B$$

where B is the background luminance.

This relationship holds over the spread of value of B encountered by pilots, the lowest of which is cockpit instrument lighting at night, and the highest dazzling fog in daylight. The corresponding values of E_t are $10^{-6.1}$ lux and $10^{-3.47}$ lux.

RVR instrumentation

Measurement of atmospheric transmission

It is possible to derive a useful visual range value only from a direct measurement of the atmospheric transmission of light, i.e. by using a transmissometer.

An instrument measuring light scattered in the atmosphere is not satisfactory. It cannot be calibrated for the wide range of atmospheric conditions that give rise to reduced visibility (the presence of fog, snow, rain and dust) and it totally ignores any contribution to light attenuation due to absorption in polluted atmospheres.

For these reasons most manufacturers have concentrated on the development of transmission measuring systems, detecting losses from both scattering and absorption processes and providing a relatively large and undisturbed atmospheric sample for the instrument.

Matching the system to the human observer

A potential problem arises with all instrumental RVR systems because the volume of the atmosphere sampled by the sensor is necessarily small when compared with the volume of atmosphere covering a runway. The validation of the use of a small sample lies in the fact that fog is produced in a dynamic situation where turbulent mixing of air is occurring.

Thus, by time-averaging it is possible to duplicate what the observer achieves with his larger spatial averaging. The validity of his conclusion is confirmed by the high degree of correlation shown between results obtained from human observers and IVR systems during the two winters of detailed evaluation.

It must be pointed out, however, that the success of this time-averaging technique is critically dependent on two additional factors: the use of a high initial sampling rate and subsequent data processing in the system computer.

Observation of fog structure has shown that large scale variations in fog density can occur from instant to instant in the sampled volume. If the sampling rate used by an instrument is too low, the mean value of fog density derived by that instrument will have a large uncertainty associated with it. For this reason, instruments using flashed sources operating at rates of the order of one pulse per second can produce indications of large short-term variations in visibility not experienced by observers. To overcome this limitation, the IVR system employs a basic sampling rate of 3.9 kHz in the transmissometer which is analogue smoothed to an output bandwidth of approximately 1 Hz.

Processing of transmission values

The analogue smoothing of the IVR transmissometer, extended by simple averaging in the system computer to give one minute mean values, enables visual range data to be produced which agrees with that from a human observer. Analysis of many thousands of hours of comparative trials between observers and the IVR system, however, has shown that both the human observer and the automatic system experience large, short-term fluctuations in visual range which are not operationally significant. Further simple averaging of RVR values would reduce this 'noise', but would introduce unacceptable delays in reporting decreasing visibility conditions. The IVR system therefore, uses a more sophisticated process based on a weighted averaging technique programmed into the system computer. The program enables a variable smoothing to be applied to RVR reporting. Rapid decreases in visibility are faithfully reported using a short 'time constant'. Operationally insignificant transient increases in visibility are suppressed, thereby increasing the usefulness of the data presented to the air traffic controller.

Computation of the visual range of lights

In addition to atmospheric transmission, it is necessary to know the value of the background luminance to calculate the visual range of a light. Considering how important this aspect of RVR calculation is to the safe operation of an airfield in fog by day by night, it is surprising how many RVR reporting systems make inadequate provision for background luminance measurement. Either a manual switch, with at most four levels that can be set by an operator, is provided, or a simple photometer is included with limited dynamic range and insufficient resolution.

From an analysis of Allard's law it is relatively easy to define the necessary requirements for a background luminance monitor such that significant errors are not introduced into the calculation of RVR. This has been done for IVR system which uses a background luminance monitor dividing the range from night to bright day into thirty steps on a logarithmic scale. This enables background luminance errors in RVR for 'worst case' conditions to be limited to less than one sixth of an ICAO RVR reporting step.

Automatic system calibration

Due to the exponential nature of the law relating to atmospheric transmission and path length through the atmosphere, difficulties are encountered in providing a wide dynamic range of visibility reporting from an instrument using a fixed measurement base line of several metres. These difficulties relate to the high accuracy required in the measurement of transmission for the assessment of visibilities considerably larger than the atmospheric sample length.

The equipment has built-in calibration devices under the control of the central processor for short-term drift correction. In addition, periods of clear weather are identified as an absolute reference standard for 100% transmission. This process relies on the interpretation of instrument voltages received from each field site to allow accurately for the effects of contamination on the external optical surfaces. The technique is made feasible only by the use of the system computer included in the central processor to which all field sites are connected.

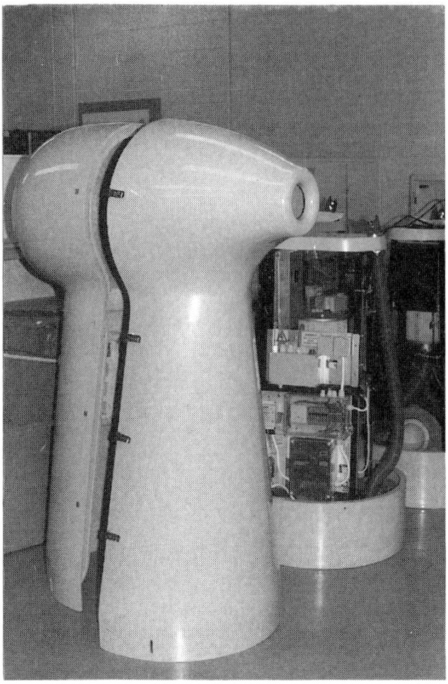

Plate 43 The protective cover for the field site electronics unit. *(Photo: Aeronautical or General Instruments Ltd)*

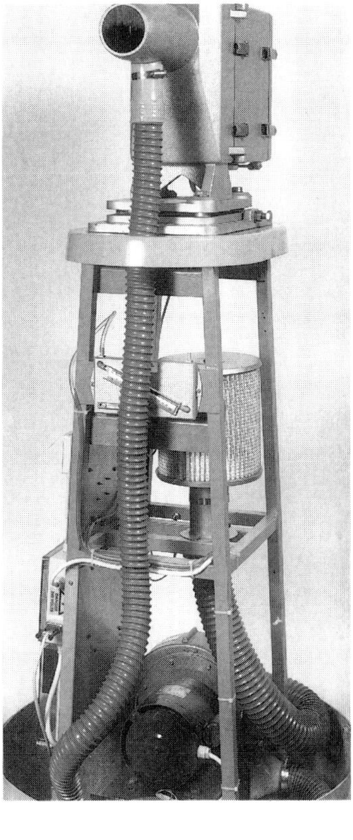

Plate 44 RVR. The field site electronics unit with the cover removed. *(Photo: Aeronautical and General Instruments Ltd)*

Plate 45 The IVR processing equipment complete with display unit, magnetic cartridge recorder and printer unit. *(Photo: Aeronautical and General Instruments Ltd)*

Elements of the IVR

A photometric measurement of atmospheric transmission (*T*) is made at each of the field sites which are mounted alongside the runway.

The measuring instrument is a transmissometer which consists of a transmit/receive unit and a tetrahedral reflector mirror, placed 10 m apart to give an optical base line of 20 m. The optical system is fully protected from dust, rain, insects, birds, etc. and the transmissometer accuracy is maintained by internal calibration devices remotely controlled by the central processor.

On one field site, there is Background Luminance Monitor (BLM) which is a digital photometer measuring Background luminance (*B*).

The analogue signals produced by the transmissometer and background luminance monitor are digitised at the field site. Digitising in the field is necessary to give a high degree of accuracy, immunity from interference and the facility of integrity checking the message. It also permits serial transmission of the data over a pair of normal speech telephone lines from the field site to and from the central processing unit. Control monitoring signals are combined in the data message.

The central processor, usually installed in the telecommunications equipment room, incorporates a general-purpose digital computer, which, in addition to receiving information from the field sites, accepts signals by wired connections indicating the brightness of runway lights, the direction of the runway in use and the time reference for the airport. From this data it performs the computation of RVR, looks after calibration and integrity checking and controls the format and logging of data.

The field site data received at the central processor are staticised into parallel data and read by the computer. The data is temporally smoothed before being used for RVR calculation at a refresh rate of thirty-six data cycles per minute.

The values of RVR and atmospheric transmission from the field sites, together with the time and other necessary particulars, are printed out by teletype whenever the RVR changes. Identical information is recorded to meet requirements for computer collation of the data with other air traffic control records.

The teletype is also used for inserting time synchronisation and other special data into the computer.

A monitoring panel on the central processor unit provides alarm indication and metering facilities.

The user's display provides a digital readout of RVR on a 15.25 cm (6 in) square panel. Upwards and downwards trends of RVR are also indicated together with runway direction and runway light intensity setting.

Components of the system

Transmissometer

The transmissometer is a photometric device for measuring the atmospheric transmission. It consists of a transmit-receive sensor unit and a reflector unit situated at opposite ends of the instrument base line.

The light source is a slightly under-run, high quality tungsten filament lamp with colour temperature similar to that of the runway lights and the spectral response of the receiver photodiode is matched to that of the CIE standard photopic observer. In this way correct weighting is given to all parts of the visible spectrum, guaranteeing that the transmission measurement relates accurately to that experienced by a human observer.

A lamp monitoring system is incorporated in the optical unit with automatic gain control (a.g.c.) in the instrumental pre-amplifier to remove the effect of residual variations in light output.

The reflector is a high-quality glass corner cube, making the instrument insensitive to changes in the angular alignment of the reflector unit.

The folded path design allows a built-in calibration reflector to be used in the transmit-receive unit to check for drift in the optical and electronic system characteristics.

The atmospheric measurement path length of 2 × 10 m has been carefully optimised so that the IVR system can accurately report RVR values from 1500 m in daylight to 50 m at night.

Background luminance monitor

Two photodetectors are used to cover the 10 000:1 dynamic range of background luminance values required. The analogue signals are compressed to the central processor. The computer converts background luminance values into eye-illuminance thresholds for use in the Allard's law calculation.

Only one background luminance monitor is normally required for each runway in an IVR installation. The unit is arranged at the field site to view a solid angle of the sky which excludes direct sunlight. The field of view excludes all ground installations and the effects of mast or airborne lights are insignificant owing to the integration over a wide field of view.

9.2 Speech and radar recording systems

The recording of radio telephony and radar signals for subsequent analysis in case of accident or incident has always been regarded as desirable if not essential.

Audio recording

Audio recording systems have been in use for many years, one of the earliest used in civil aviation being the Simon recorder. This equipment utilised a recording head somewhat similar to that of a disc recorder which impressed a track on a loop of plain 35 mm cine film. The loop was approximately 80 ft long and included a twist such that recording was possible on both sides of the film. As the recording progressed about eighty tracks were impressed across the useable width of the film, this giving a duration of about eight hours.

Two separate recording decks had to be provided for each channel which, together with the automatic changeover equipment were mounted in a single equipment rack. This system met the dual requirement of long duration and permanency of recording but required extreme care on the part of technical staff to ensure freedom from tape breakage and reasonable recording quality.

The Simon recorder remained in service for many years but was superseded in the early 1960s by multi-channel magnetic tape recording.

In this equipment up to thirty-two tracks could be recorded on a single one inch wide tape and separate tracks could be allocated for continuous time injection and engineering purposes. A further track was designated as a 'spare'.

Separate heads were used for recording and replay, thus continuous monitoring of recorded signals was possible. An automatic monitoring system was also incorporated which arranged for a tone to be recorded on each track and monitored at the replay head. After detection, further circuitry filtered out the tone in the replay circuits. If a faulty channel was detected, service was automatically switched to a spare channel. In the event of a second channel failure, the recorder was switched off and service transferred to stand-by equipment.

Such equipment has been produced by several manufacturers, the type of equipment installed at any given location being largely determined by channel capacity requirements. Such capacity varies from eight channels on quarter inch tape to thirty-two or forty-eight channels on one inch tape. Tape speed is normally $\frac{15}{16}$ inches per second giving a standard 2400 foot reel of tape a duration of eight hours. Shortly before the end of a tape is reached, a stand-by recorder is energised and both operate in parallel until tape run-out on the first recorder. Service is continued by the second recorder, the first recorder reverting to stand-by or reserve status.

Volmet broadcasts

Recording systems are widely used for Volmet broadcasts. Each of these is a continually repeating transmission which comprises a number of reports of weather conditions at airports within the region concerned. As, however, weather conditions are rarely static, provision has to be made for updating each station report individually.

In the earliest equipment, the report for each station was recorded on a loop of tape, with the loops being replayed in turn to make the complete broadcast. When an amended report was received, so the corresponding loop was re-recorded. This system was reasonably effective and still remains in service in some parts of the world. It suffered, however, from poor audio quality due to tape wear, and from the fact that one transmission may include contributions from several operators, each with a different regionai accent. The first of these problems has been overcome in many centres by digitally recording the transmission and retaining it in a solid state memory, thus eliminating any tape noise, etc.

A more advanced computerised system has, however, been introduced by Marconi Secure Radio Systems Ltd, which eliminates the operator and fully

Plate 46 A Marconi Secure Radio Systems Automatic Volmet equipment. *(Photo: Marconi)*

automates the system. Meteorological reports are originated at airports and passed to the Volmet station by teleprinter using a standard format. The first task of the computer is to receive the incoming message and check its contents for errors, the second is to decode the message into plain text and the third to assemble the spoken broadcast by selecting the equivalent individual words from the spoken vocabulary held in the digital store.

The vocabulary software was developed by recording each word in its context in a typical sentence. The word in digital form was then 'cut out' using a computer working to an accuracy of a few milliseconds and fitted by an iterative technique to every other word with which it may be associated.

Words which might occur both in the middle and at the end of a sentence were recorded twice, with appropriate cadence. The software also contains several periods of silence for use in various contexts.

A male voice was chosen for these recordings as it permits greater fidelity for less digital storage. The speaker was chosen for the clearness, consistency and lack of a regional accent in his voice, thus making it easier for people from other countries to understand.

The avoidance of human error is one of the chief advantages of automation. Errors could occur in non-automated systems either in the decoding or recording on the ground, and in hearing the broadcast incorrectly in the air. From the point of view of the pilot the familiarity of the same voice, in addition to its clarity and consistency, is a valuable aid to intelligibility.

A final advantage is that the computer updates the broadcast immediately a new report is received, thus gaining at least ten minutes over manual systems.

Radar recording

Although the recording of speech has been in use for many years, the recording of radar has proved more difficult. Analogue radar signals cover a very broad bandwidth and to ensure satisfactory reproduction very high tape speeds are necessary which make the use of long duration tapes impractical. The introduction of plot-extracted radar greatly reduced the bandwidth requirements thus permitting the use of standard digital instrumentation recorders. Such techniques were introduced at London Air Traffic Control Centre in early 1978; however, these were superseded in 1990 by a digital recording system. Nevertheless, a description of the earlier system is worthwhile for it is typical of many which were, and are still, in use in many parts of the world.

The system at London air traffic control centre incorporated three SE Labs' SE 7000 series portable analogue tape recorders. These were high precision IRIG standard instruments which were sequenced by an automatic control and monitoring unit in such a way that one was always in the record mode, and another was on record stand-by, whilst the third was used for replay or as a back up to the other two.

Using any of the three recorders in the 'playback' mode, it was possible to generate a dynamic display of aircraft movements on a radar screen for subsequent analysis. Print-outs of individual aircraft tracks could also be obtained by using suitable computer programs in the CAA radar processors.

Data recording

The digital information recorded included both primary radar data (aircraft position data generated by surveillance radar) and secondary radar data; the latter being produced by SSR transponder equipped aircraft and including such information as aircraft position, identity and flight level.

The primary and secondary digitised radar data fed to the system was fully plot extracted. This was received at 7200 b/s with thirteen-bit fields and an idle field present after each target message. Secondary only radar data was input at 3600 b/s with eighteen-bit fields and idle fields generated when no target messages were available. The other external inputs to the system were time codes (in IRIG B–days, hours, minutes, seconds) using a 1 KHz carrier modulated by serial bed time, and a voice input on each recorder using a handset. The main internal data inputs to the system were servo signals used for tape checking routines.

The monitoring facilities checked and indicated the continuous presence of

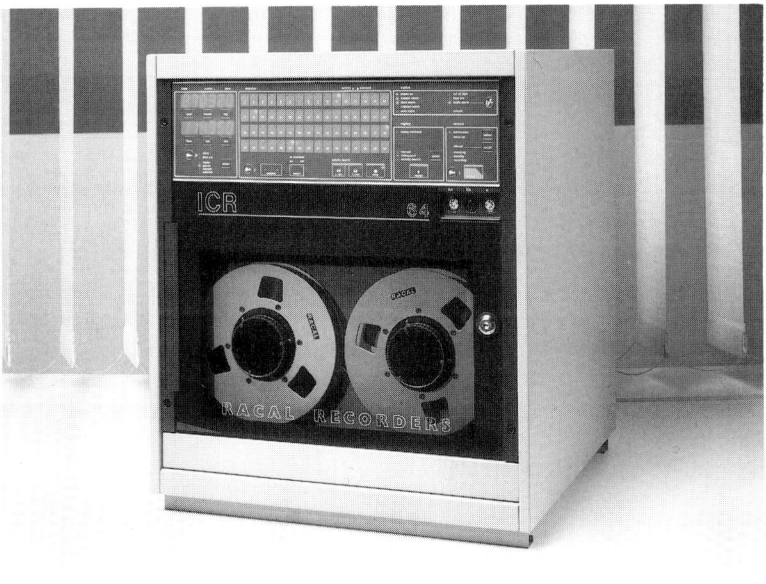

Plate 47 Racal Recorders' ICR64 communications recording system, typically used in air traffic control centres. *(Photo: Racal Recorders Ltd)*

an input signal and a correct output signal on all tracks except voice; the status of all three recorders (on-line, stand-by reserve, playback or out of service); the need for a tape change; any fault indications e.g. input and record failure; and staticised fault indications on a previously on-line machine. Monitoring of the recorded radar data track outputs was carried out by digital check circuits which detected idle fields present in the radar data.

The control system automatically sequenced the three recorders – the outgoing machine continued recording until the new on-line machine was checked out – re-spooled recorded tapes, and indicated that a tape change was required. This reduced routine operations to tape loading and occasionally cleaning the tape guides, rollers and headstacks. A check-tape routine ensured that previously recorded tape could not be overwritten, and further system integrity was provided by inhibiting recordings within twenty feet of both ends of each tape.

High quality back coated instrumentation tape was used, a 4600 ft reel of which had a sixteen hour capacity at $\frac{15}{16}$ inches per second. Tape run time on each recorder could be individually varied from seven and a half minutes up to the limit imposed by tape length. Tape utilisation was twenty-three tracks of radar data (up to 7200 b/s); two tracks each of time reference data and data protection signals, and one track for voice recording when required.

A standard feature of this type of recorder was that odd and even tracks were split within the machine, e.g. the record head was split into two fourteen-track blocks, one even, one odd, which was interleaved on the recorded tape. This concept was carried through the entire system, and in the LATCC application each radar channel consisted of two identical radar data inputs: one was recorded on an odd numbered track, and the other on an even numbered track, for enhanced reliability.

Recordings from both audio and video equipment were retained for one month after which, unless required for incident or accident investigation, the tapes were erased for re-use.

Digital radar recording

The pace of development of digital electronics made it inevitable that the analogue systems used for both audio and radar recording should be replaced at the earliest opportunity. Consequently, the method of radar recording outlined above was replaced at London Air Traffic Control Centre in mid-1990 by a fully digital system.

The system chosen was the 'Hindsight' equipment developed by Walton Radar Systems, and this provides an excellent introduction to the new generation of combined radar and audio recorders.

Although 'Hindsight' has capability for both radar and audio recording, the latter function was not incorporated in the London Air Traffic Control installation.

Plate 48 Hindsight recorders. The local control unit is at the top of each rack below which is the Mass Storage Unit; System Control Unit; data interface unit. *(Photo: Walton Radar Systems)*

Basic principles

The incoming data are recorded on a single track on 8 mm cassette tape. In order to achieve a high data storage, helical scanning is used in a similar manner to a domestic video recorder.

One problem is that only one track is available for recording perhaps 40 different radar or audio channels. The capability must therefore be available to identify which information is appropriate to which channel. This is

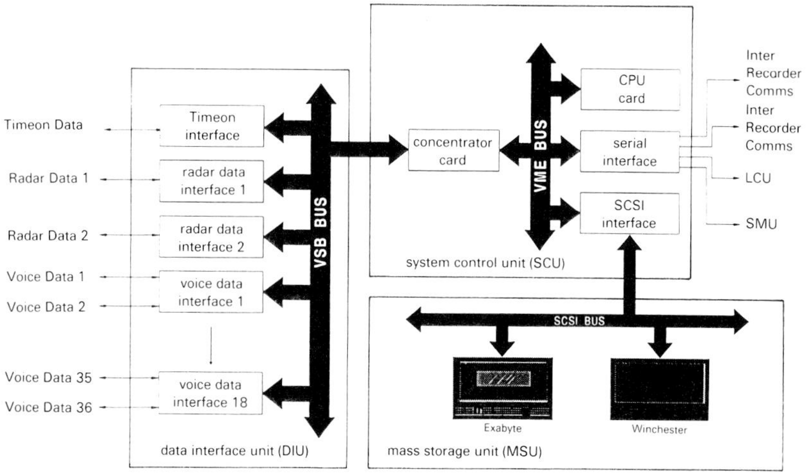

Fig. 62 Block diagram of the 'Hindsight' radar and voice recording system. *(Photo: Walton Radar Systems)*

facilitated by the nature of the plot extracted radar signals, for these comprise digital words describing the position of each target. Prior to recording, a preamble is added to each word giving both channel identification and timing.

Where audio is being recorded, the analogue audio signal is first digitised, then divided into short 'packets' which are then handled in exactly the same manner as the radar signals.

On replay, the software identifies the signals appropriate to the channel of interest; these are then processed prior to being fed to the radar display unit or audio system as appropriate.

The 'Hindsight' system

A major installation such as London Air Traffic Control Centre comprises three separate recorders; at any one time, one will be on service, a second in standby mode and the third acting as a reserve or under maintenance.

The data to be recorded are 'daisy chained' to all three recorders simultaneously where they first enter the Data Interface Unit (DIU). Here the input data, which may be in any of a number of different formats, are reformatted and output to a VSB bus. Audio inputs are digitised, compressed by a digital signal processor and then stored in the common data collection bus interface buffers.

A central processing unit (CPU) located in system control unit (SCU) controls the output from all interface units which is then fed through a concentrator card which compresses the data by a factor of up to ten times the normal density without affecting the replay.

The data buffers on the concentrator card are monitored by the CPU which then instructs the Small Computer Systems Interface (SCSI) card to collect the data and route it to the appropriate mass storage device via a SCSI bus.

The Mass Storage Unit (MSU) contains two Exabyte cassette drives and a high capacity Winchester disk. In operational mode, the incoming data from the SCSI is stored directly on the Exabyte cassettes with the Winchester only storing system configuration information so that, in the event of power failure, system reconfiguration is not necessary.

When in standby or reserve mode, however, the cassette drives are inactive and all incoming information is recorded on the Winchester disk.

Changeover

Should the on-line recorder fail or be taken off-line before the scheduled time, the Winchester disk in the standby recorder is used to 'catch up' with events.

When this occurs, current data continue to be recorded on the Winchester disk but those already on the disk are transferred to the cassettes. However, as the disk-to-cassette transfer rate is much higher than that of the incoming data, the amount held on the disk quickly diminishes; and when this falls to zero, direct recording on cassette can resume.

Data replay

The replay process is basically the reverse of that for recording except that the data are always transferred from tape to disk and then read from the disk to the output device.

As each packet of information on the tape is time coded, a timesearch is possible to enable the replay to commence a little before the incident to be investigated.

A further facility offered by the 'Hindsight' system is for rapid replay when the recorded data can be replayed faster than they actually arrived but without increasing the data rate.

This is achieved by reading the data from one complete antenna revolution and then skipping a number of revolutions. 'Skip rates' of 2, 4, 8 and 16 are immediately available.

Immediate replay

In the past, when an emergency occurred, the tape was immediately removed from the recording apparatus for analysis. In the 'Hindsight' system, however, this is not necessary, for all incoming data are continually recorded on the Winchester disks of the non-operational recorders. In an emergency the D & D (Distress and Diversion) hold button on the Remote Status Unit can be pressed which will 'freeze' the data on the lowest priority recorder. From this time, that recorder will take no further part in the recording cycle. This machine will then be available for immediate replay of the incident.

Appendix 1

Units of measurement used in telecommunications

Quantity	Name	Symbol
Potential difference or electromotive force	Volt	V
Electrical current	Ampere	A
Electrical resistance	Ohm	Ω
Time	Second	s
Electrical capacitance	Farad	F*
Frequency	Hertz (1 Hz = 1 cycle/s)	Hz
Inductance	Henry	H
Power	Watt	W
Relative power of voltage	bel	B**

*Most commonly μF or pF.
**Most commonly dB. Note also dBm which is power relative to 1 mW.

Multiples and sub-multiples

Factor by which the unit is multiplied	Prefix	Symbol
1 000 000 000 = 10^9	giga	G
1 000 000 = 10^6	mega	M
1 000 = 10^3	kilo	k
0.1 = 10^{-1}	deci	d
0.001 = 10^{-3}	milli	m
0.000 001 = 10^{-6}	micro	μ
0.000 000 001 = 10^{-9}	nano	n
0.000 000 000 001 = 10^{-12}	pico	p

Appendix 2

Designation of radio emissions

The 1979 World Administrative Radio Conference adopted a proposal for a new convention to be used world-wide for the designation of the characteristics of each individual type of radio emission. This was brought into use on 1 January 1982.

The designation comprises three symbols, of which the first describes the modulation of the main carrier, the second describes the nature of the modulating signals and the third, the type of information transmitted.

First Symbol

Type of modulation of the main carrier

N Emission of an unmodulated carrier

Emission in which the main carrier is amplitude-modulated (including cases where sub-carriers are angle modulated)

A Double-sideband
H Single-sideband, full carrier
R Single-sideband, reduced or variable level carrier
J Single sideband, suppressed carrier
B Independent sideband
C Vestigial sideband

Emission in which the main carrier is angle-modulated

F Frequency modulation
G Phase modulation
D Emission in which the main carrier is amplitude and angle-modulated either simultaneously or in a pre-established sequence

Emission of pulses*

P Unmodulated sequence of pulses
K A sequence of pulses modulated in amplitude
L A sequence of pulses modulated in width/duration
M A sequence of pulses modulated in position/phase
Q A sequence of pulses in which the carrier is angle-modulated during the period of the pulse
V A sequence of pulses which is a combination of the foregoing or is produced by other means

* emissions where the main carrier is directly modulated by a signal which has been coded into quantised form (e.g. pulse code modulation) should be designated under amplitude- or angle-modulation as appropriate

W Cases not covered above in which an emission consists of the main carrier modulated either simultaneously or in a pre-established sequence in a combination of two or more of the following modes: amplitude, angle, pulse
X Cases not otherwise covered

Second Symbol

Nature of signal(s) modulating the main carrier

0 No modulating signal
1 A single channel containing quantised or digital information without the use of a modulating sub-carrier†
2 A single channel containing quantised or digital information with the use of a modulating sub-carrier†
3 A single channel containing analogue information
7 Two or more channels containing quantised or digital information
8 Two or more channels containing analogue information
9 Composite system with one or more channels containing quantised or digital information, together with one or more channels containing analogue information
X Cases not otherwise covered

† This excludes time-division multiplex

Third Symbol

Type of information to be transmitted‡

N No information transmitted
A Telegraphy – for aural reception
B Telegraphy – for automatic reception
C Facsimile
D Data transmission, telemetry, telecommand
E Telephony (including sound broadcasting)
F Television (video)
W Combination of any of the above
X Cases not otherwise covered

‡ In this context the word "information" does not include information of a constant unvarying nature such as provided by standard frequency emissions, continuous wave and pulse radars, etc.

Examples

Single sideband, suppressed carrier	J3E
Amplitude – modulated vestigial sideband television	C3F

Appendix 3

Radio frequency band designations

	Old		New
P	80–390 MHz	A	0–250 MHz
L	390–2500 MHz	B	250–500 MHz
S	2.5–4.1 GHz	C	500–1000 MHz
C	4.1–7.0 GHz	D	1–2 GHz
X	7–11.5 GHz	E	2–3 GHz
J	11.5–18 GHz	F	3–4 GHz
K	18–33 GHz	G	4–6 GHz
Q	33–40 GHz	H	6–8 GHz
O	40–60 GHz	I	8–10 GHz
V	60–90 GHz	J	10–20 GHz
		K	20–40 GHz
		L	40–60 GHz
		M	60–100 GHz

Note: In order to define a channel to closer limits, each band is subdivided into ten equal channels, numbered 1–10, starting at the lower frequency limit (i.e., 1.0 to 1.1 GHz is channel D1).

To specify frequency more precisely, quote the channel number and then add the number of megahertz necessary, counted from the lowest channel frequency. (i.e. 260 MHz is B1 plus 10).

To avoid confusion, the upper frequency limit of letter designators is considered to be 'up to but not including'.

Glossary

a.c.	Alternating current.
ADF	Automatic Direction Finder. Usually refers to airborne equipment operating in the 200 kHz to 500 kHz band.
Adsel	Address Selective SSR. Recent development of SSR which permits interrogation of individual aircraft and exchange of data. Compatible with USA Discrete Address Beacon System (DABS).
af	Audio frequency. Normally refers to frequencies between 20 Hz and 25 000 Hz.
AFC	Automatic Frequency Control. Error correcting circuit which may be fitted to receiving equipment to compensate for any frequency variation in either the receiver circuits or the received signal.
AFTN	Aeronautical Fixed Telecommunication Network. A communications network connecting all principal airports and air traffic control centres intended for the dissemination of flight plans, NOTAMS etc.
AM	Amplitude Modulation. The act of impressing speech or other data on a radio frequency signal by varying the instantaneous power level.
Ampere	The unit of electrical current.
A-scope	A type of radar display in which range is indicated by displacement along the trace and received signals cause a deflection at right angles to the trace.
ATCC	Air Traffic Control Centre.
ATCO	Air Traffic Control Officer. A person trained and licenced to control air traffic.
Bellini-Tosi	Bellini and Tosi were two early experimenters who developed the direction finding system named after them. This determined the direction of the incoming signal by comparing the relative signal strengths received on two fixed loop aerials mounted at right angles.
bcd	Binary coded decimal.
bco	Binary coded octal.
BFO	Beat Frequency Oscillator. (a) An oscillator in a receiver, tuned to intermediate frequency which will either beat with incoming cw signals or substitute for the carrier of SSB signals before detection. (b) An early form of wide range audio oscillator which

mixed the output of two RF oscillators, one fixed and one variable, to obtain an audio output.

BLM	Background Luminance Monitor. Used in conjunction with transmissometers in IRVR systems.
b/s	Bits per second. A measure of rate of passage of digital data.
CAA	Civil Aviation Authority.
CCTV	Closed Circuit Television in which the interconnection between camera and display is by electrical cable or optical fibre.
CIE	Commission International de l'Éclairage (International Commission on Illumination).
Clarifier	Alternative nomenclature for BFO tuning control on some receivers designed primarily for SSB reception.
Control area	A volume of airspace whose geographical area and upper and lower limits are notified, in which air traffic control exists.
COSPAS	The Soviet equivalent of the SARSAT satellites, which works in conjunction with them.
CRDF	An automatic direction finder, usually VHF or UHF, which incorporates a cathode ray tube for bearing display.
CRT	Cathode Ray Tube. An evacuated glass bulb in which a beam of electrons emitted by a heated cathode, focussed by a number of anodes and deflected by either electrostatic or electromagnetic means, impinge on a screen coated by various phosphors. This action causes an observable fluorescence of the phosphors.
CSB	Carrier and Sidebands (ILS terminology).
cw	Continuous wave. An RF signal modulated by on-off keying.
dB	Decibel. The unit of relative power or voltage, measured on a logarithmic scale.
dBm	The relative unit of power compared with a standard of 1 mW across an impedence of 600 ohms.
d.c.	Direct current.
ddm	Difference in depth of modulation. The comparison of the modulation depths of the 90 Hz and 150 Hz tones. (ILS terminology.)
DFTI	Distance From Threshold Indicator. A specialised PPI radar display, used normally in visual control rooms, which displays the segment of the radar signal corresponding to the final approach path of the runway in use.

Dipole	An aerial which is split and fed at its centre point.
DME	Distance Measuring Equipment. The distance measuring element of VOR/DME, the standard short range navigational aid. This equipment uses secondary radar principles and operates in L-band.
Doppler effect	The apparent change in frequency of a wave motion observed, due to relative motion between the source of radiation and the observer.
DVST	Direct Viewing Storage Tube. A type of cathode ray tube, which by use of storage techniques, enables the picture to be displayed at high intensity.
ELT	Emergency Locator Transmitter.
EPIRB	Emergency Portable Indicating Radio Beacon.
FDPS	Flight Data Processing System.
fet	Field effect transistor.
FIR	Flight Information Region. A geographical area under the jurisdiction of a single air traffic control centre.
Flight level (FL)	When flying with the altimeter set at 1013.2 mb, heights are measured in hundreds of feet and are referred to as flight levels.
FM	Frequency Modulation. The act of impressing speech or other data on a radio frequency signal by varying the transmitted frequency.
FOG	Fibre-optic gyro.
FPPS	Flight Plan Processing System.
Fruit	Non-synchronous SSR replies.
Glide path	The approach path, in the vertical plane, to a runway.
Glonass	A global positioning system developed by the former Soviet Union.
GMC	Ground Movements Control. A common term for the air traffic control position which controls aircraft taxiing and some vehicular movements.
GMS	Geostationary Meteorological Satellite.
GPS	Global Positioning System. A world-wide system by which a user may determine his position by the comparison of signals from two or more satellites.
Hz	Hertz. The unit of frequency. (1 Hz = 1 cycle per second.)
IC	Integrated Circuit.
ICAO	International Civil Aviation Organisation.

IF	Intermediate Frequency.
ILS	Instrument Landing System. The standard ICAO approach aid.
INS	Inertial Navigation System. A navigation system in which displacement from the point of departure is determined by measuring the acceleration exerted upon a gyroscopically stabilised platform by vehicle movement.
Interscan	(a) The period between the end of one scan of a timebase on a radar display and the commencement of the next. (b) The name of the Australian contender for the MLS international standard. This operates on TRSB principles and the proposal was eventually combined with that from the USA to form the preferred standard.
IMC	Instrument Meteorological Conditions. Meteorological conditions, below certain predetermined criteria, in which Instrument Flight rules apply.
IRIG	An acronym for the Inter Range Instrumentation Group affiliated to the United States Dept of Defense. In 1948, this group set standards for instrumentation recording, which have been revised periodically and are still recognised today.
ISB	Independent Side Band. A system of radio transmission in which two single sideband transmissions, one upper sideband and lower sideband, both associated with the same carrier frequency, carry independent information.
IVR	Instrumental Visual Range – an equipment providing an automatic method of assessing visibility.
Knot	The unit of velocity used in aviation. (1 knot = 1 nautical mile per hour).
LATCC	London Air Traffic Control Centre.
LORAN A	LOng Range Aid to Navigation. Originally introduced in the 1940s, this pulse hyperbolic system operated on frequencies in the 1.9 MHz band. Now obsolete.
LORAN C	A replacement system for LORAN A which uses pulse hyperbolic techniques and operates on 100 kHz.
LORAN D	A shorter range version of LORAN C.
LSB	Lower Sideband. The sideband of an AM transmission which is of lower-frequency than the carrier.
Magnetron	A high power transmitting valve frequently used in radar transmitters.
MCW	A type of AM modulation in which the information is imparted by on-off keying of a modulating tone.

Meteosat	The European Geostationary Meteorological Satellite
Microwaves	Term used to describe all frequencies above 1000 MHz.
MLS	Microwave Landing System. The proposed replacement for ILS.
Mosaic	A system of radar processing in which the outputs of several radar units at differing locations are combined to provide a single composite picture.
MOTNE	Meteorological Operations Telecommunications Network Europe. A teleprinter network for disseminating meteorological information throughout Europe.
MTD	Moving Target Detector. The circuitry concerned with identifying valid primary radar returns while eliminating permanent echoes and other unwanted signals by use of digital techniques.
MTI	Moving Target Indicator. The circuitry concerned with identifying valid primary radar returns while eliminating permanent echoes and other unwanted returns by use of analogue techniques.
MTM	Multiple Transistor Module. A means of obtaining high power output from solid state radar transmitters. The final output level is obtained by combining the output from many low modules.
NDB	Non-Directional Beacon. A radio transmitter, operating on the 200 kHz – 500 kHz band identifying a geographical location which may be located by airborne ADF equipment.
nm	Nautical miles.
Ohm	The unit of electrical resistance.
Omega	A long range hyperbolic navigational system which operates on the VLF band.
OSCAR	Orbiting Satellite Carrying Amateur Radio.
PA	Power Amplifier.
PEP	Peak Envelope Power.
Polar diagram	A graphical method of indicating the directional characteristics of an aerial system in a single plane. If that plane is vertical the diagram is called a Vertical Polar Diagram (VPD) and if horizontal, a Horizontal Polar Diagram (HPD).
Plot extraction	The process of identifying a valid radar return and converting the positional information from Rho/Theta (i.e. Range/Bearing) to cartesian coordinates delineated by a digital word.

PM	Phase Modulation.
PPI	Plan Position Indicator. A form of radar display which indicates the relative position of targets in the horizontal plane.
PRDS	Processed Radar Display System. The system used at LATCC which accepts radar data from remote radar stations, applies suitable processing and displays the signals in suitable form for use by ATCOs.
PRF	Pulse Recurrence Frequency. The number of pulses or pulse groups radiated per second by a radar or other aid which uses pulse techniques.
Q-code	An international telegraph code, originally intended to accelerate commercial WT operation which has since proved so convenient that it is also used in RT operation. Typical examples: QFE The atmospheric pressure at airfield levels is . . . QTE Your true bearing from this station is . . .
Q-Band	Original nomenclature for the 35000 MHz (8 mm) microwave band (see Appendix 3).
RDPS	Radar Data Processing System.
RMI	Radio Magnetic Indicator. An instrument which is installed in an aircraft for indicating a bearing derived from either VOR or ADF equipment.
R/T	Radio Telephony.
Rhumbatron	A cavity, which by virtue of its dimensions acts as a resonant tuned circuit. Used on microwave equipment.
RVR	Runway Visual Range i.e. visibility along the runway. May be assessed manually or by instruments in which case it is described as IRVR.
Rho-Theta	Delineation of position by specifying range and bearing from a fixed point.
SARSAT	Search And Rescue Satellite.
S-Band	Original nomenclature for frequencies in the order of 3000 MHz (see Appendix 3).
SBO	Sidebands only (ILS terminology).
Scan-conversion	The process of converting radar video signals from Rho/Theta to TV type radar display.
SCS 51	The original ILS system.
SGC	Swept Gain Control, also known as Sensitivity Time Control (STC). The action by which a radar receiver is desensitised immediately after the transmitter fires, but gradually restores so as to maintain all signals at a reasonably constant signal strength at the detector.

SHF	Super High Frequencies 3 GHz–30 GHz.
SLS	Side Lobe Suppression. The technique used in SSR to ensure that aircraft are not interrogated by signals other than those radiated by the main beam of the aerial array.
SPI	Special Position Indicator. An additional pulse radiated after the main SSR reply group which causes an indication to appear on the radar display. It is used for identification.
SRA	Surveillance Radar Approach. A technique for ground controlled approach guidance using only PPI radar display.
SSB	Single Sideband. A form of amplitude modulated transmission in which the carrier wave and one sideband are not radiated.
SSR	Secondary Surveillance Radar. A radar technique in which the ground transmission is received by the aircraft, automatically initiating a reply which can contain information such as aircraft identity and height.
Transmissometer	A component part of an instrumented runway visual range equipment which measures the attenuation due to mist or fog of a beam of light traversing a fixed path.
TMA	A confluence of airways in vicinity of an airport or airports.
TRSB	Time referenced Scanning Beam. The USA proposal to the AWO meeting in April 1978 and adopted as the future MLS technique.
UHF	Ultra High Frequency. A general term for frequencies between 200 MHz and 1000 MHz.
USB	Upper Sideband. The sideband of an AM transmission which is higher than the carrier frequency.
UTC	Coordinated Universal Time.
VCR	Visual Control Room. The room situated on the roof of the air traffic control tower, occupied usually by the tower and ground movements controllers.
VHF	Very High Frequency. A general term for radio frequencies between 30 MHz and 200 MHz.
VLF	Very Low Frequencies. A general term for radio frequencies below 100 kHz.
VMC	Visual Meteorological Conditions. Meteorological conditions which are better than certain predetermined criteria in which aircraft may fly in accordance with visual flight rules.

VOR	VHF Omni Range. The azimuth element of VOR–DME which is the international standard short range navigational aid.
Volt	The unit of electrical pressure.
VOX	Voice Operated Transmit. A circuit, sometimes fitted to radio transmitters, which switches the equipment automatically from receive to transmit when the microphone picks up a sound above a predetermined level.
X-Band	Original nomenclature for frequencies in the order of 10000 MHz (see Appendix 3).

Index